Quality of
Irrigation Water

IIIC Publication No. 2

Quality of Irrigation Water

I. Shainberg

Institute of Soils and Water
Agricultural Research Organization
Bet Dagan, Israel

J. D. Oster

U.S. Salinity Laboratory
Agricultural Research Service
Riverside, California, United States

International Irrigation Information Center
Volcani Center, POB 49, Bet Dagan, Israel
POB 8500, Ottawa K1G 3H9 Canada

DISTRIBUTED BY
Pergamon Press Ltd., Headington Hill Hall, Oxford OX3 0BW.
England
Pergamon Press Inc., Maxwell House, Fairview Park, Elmsford, New York 10523, U.S.A.
Pergamon of Canada, Suite 104, 150 Consumers Road, Willowdale, Ontario M2J1P9, Canada
Pergamon Press (Aust.) Pty. Ltd., P.O. Box 544, Potts Point, N.S.W. 2011, Australia
Pergamon Press SARL, 24 rue des Ecoles, 75240 Paris, Cedex 05, France
Pergamon Press GmbH, 6242 Kronberg, Taunus, Pferdstrasse 1, Federal Republic of Germany

British Library Cataloguing in Publication Data

Shainberg, I
Quality of irrigation water. – (International Irrigation Information Center. IIIC publications; no. 2).
I. Irrigation water – Quality
I. Title II. Oster, J.D. III. Series
628'.72   S618.45   79- 40000
ISBN 0-08-023822-X

Printed in Israel

Published 1978 by the
International Irrigation Information Center
ISRAEL: Volcani Center, P.O.B. 49, Bet Dagan
CANADA: P.O.B. 8500, Ottawa, K1G 3H9

ISBN: 92-9019-002-9

© International Irrigation Information Center, Bet Dagan, Israel

Typeset by The Israel Economist, P.O.Box 7052, Jerusalem

Cover designed by Avraham Pladot AGAF

Table of Contents

I. Introduction

In many parts of the world, available soil moisture derived from rain or from underground waters is not sufficient for the requirements of plant life, at least through part of the growing season. Such deficiency can be made up by irrigation.

Although irrigation has been practised in the world for several millennia, it is only in this century that the importance of the quality of irrigation water has been recognized. The use of saline water may result in reduction of crop yields, while sodic water may cause deterioration in the physical properties of soils, again with consequent reduction in yield. Considerable attention is given at present to the environmental aspects of water quality, including the possible presence of minute amounts of potentially harmful substances.

Quality of irrigation water is of particular importance in arid climates. Salts formed *in situ* by weathering of soil minerals or by salt deposition from applied water tend to accumulate in the soil profile.

Water used for irrigation may contain up to 3000 g of salt per cubic meter as against 5-40 g in a cubic meter of rain water. Application of 100 mm of irrigation water containing 1000 g per cubic meter introduces one ton of salt to a hectare of land. Consequently, it is necessary to include leaching and drainage as an integral part of the irrigation program, in order to maintain salinity levels compatible with continued cropping.

As water resources become more limited, increased use is being made of inferior irrigation water, with a high sodium or total salt content or both. Without proper management, based on knowledge of the possible harmful effects, prolonged application of such water is impossible.

II. Properties of Irrigation Water

The quality of irrigation water is determined by the following chemical characteristics:

1. Total concentration of soluble salts, or salinity.

2. Concentration of sodium relative to other cations, or sodicity.

3. Anionic composition of the water, especially concentration of bicarbonate and carbonate anions.

4. Concentration of boron or other elements that may be toxic to plant growth.

A detailed description of the techniques commonly employed for the analysis of irrigation water is available (Richards, 1954; Black, 1965); only the general principles are presented here.

Total salt concentration

The effect of salt on crop growth is believed to be largely of osmotic nature. Osmotic pressure is a colligative property, and is therefore related to the total salt concentration rather than the concentration of individual ionic species. Two methods for determination of salt concentration are in common use:

1. Total Dissolved Solids (*TDS*). An aliquot of a filtered solution is evaporated to dryness in a weighed platinum dish. The weight of the residue is determined and expressed in milligrams of salt per liter of

solution or parts per million (ppm). This method is accurate in the case of water free of bicarbonate. About half of the bicarbonate is lost in the process of heating, which reduces the values recorded for *TDS*.

2. Soluble Salts by Electrical Conductivity (*EC*). The *EC* of a water sample is determined by measuring the electrical resistance between two parallel electrodes immersed in the solution. Pure water is a very poor conductor of electric current, whereas water containing salt conducts an alternating current approximately in proportion to the amount of salt present. The specific conductivity, or the conductance per unit cross-sectional area and across unit distance, is obtained from the resistance recorded across a conductivity cell from the relation

$$EC = k/R \qquad\qquad\qquad\qquad (II.1)$$

where k is the cell constant and R is the resistance.

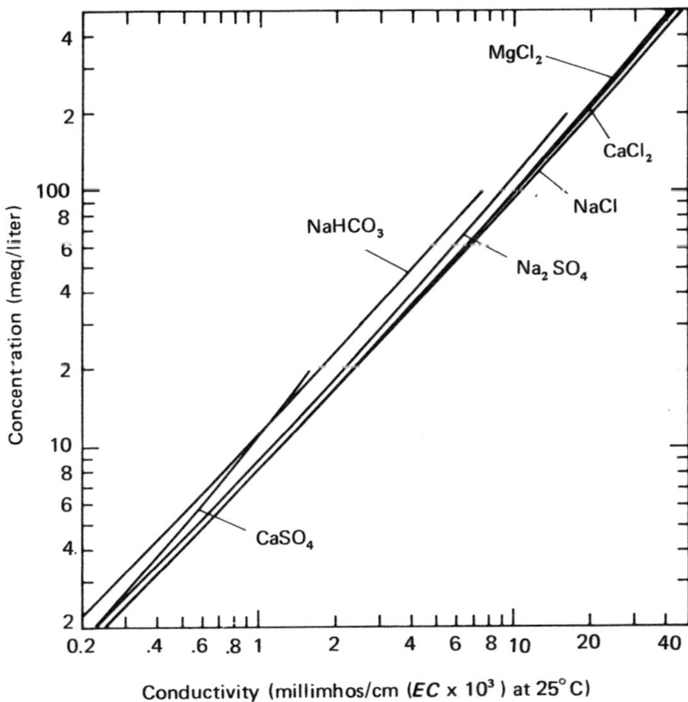

Figure II.1 Relationship between salt concentration (meq/liter) and electrical conductivity of various salt solutions.

The basic unit for EC is mho·m^{-1} or, in SI units, Siemens/meter ($S·m^{-1}$). The mho is the inverse of the ohm. Historically, it has been found convenient to use the units mho/cm (mho·cm^{-1}), millimho/cm (mmho·cm^{-1}) and micromho/cm (μmho·cm^{-1}). 1 mho·cm^{-1} is equal to 100 mho·m^{-1} or 100 Siemens/meter. Thus, a typical irrigation water may have EC = 2000 μmho·cm^{-1} = 2 mmho·cm^{-1} = 0.2 S·m^{-1}. The convenient mmho/cm unit is used most frequently in this review.

The relationships between EC and concentration (in meq/liter and in mg/liter) of various salt solutions are presented in Figs. II.1 and II.2. It is evident that when the concentration is given in mg/liter (ppm) (Fig. II.2), the curves are more widely separated than when the concentration is expressed in meq/liter (Fig. II.1).

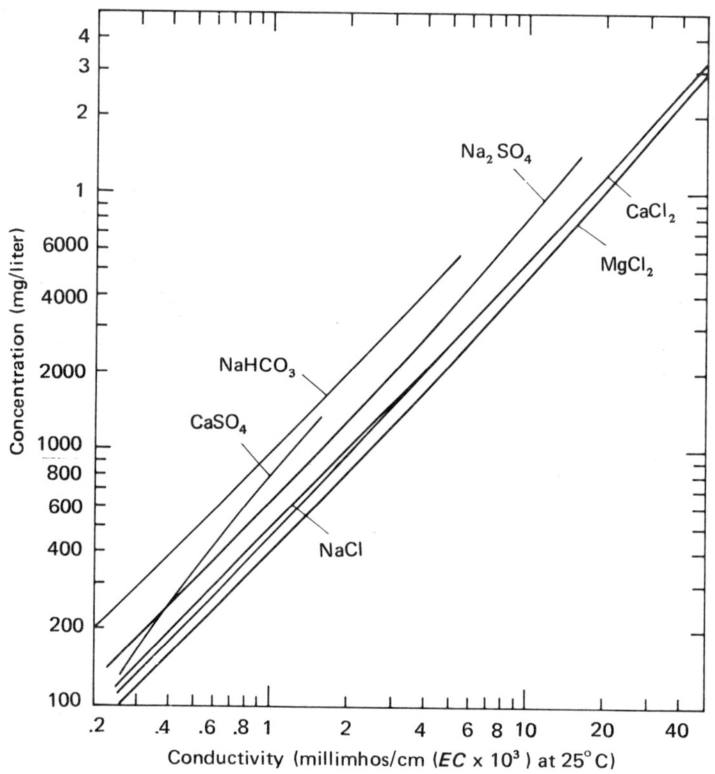

Figure II.2 Relationship between salt concentration (mg/liter) and electrical conductivity of various salt solutions.

This is because EC depends on the number of ions in the solution and not on their weight, whereas concentration in ppm depends on both. Nevertheless, for a mixture of salts in the range up to 10 mmho·cm^{-1}, there is a linear relationship between the *EC* of the water and the *TDS*:

$$TDS \text{ (ppm)} = 640 \times EC \text{ (mmho/cm)} \qquad \text{(II.2)}$$

The relationship between the osmotic pressure (*OP*) of soil extracts which involve several ions and their electrical conductivity is shown in Fig. II.3.

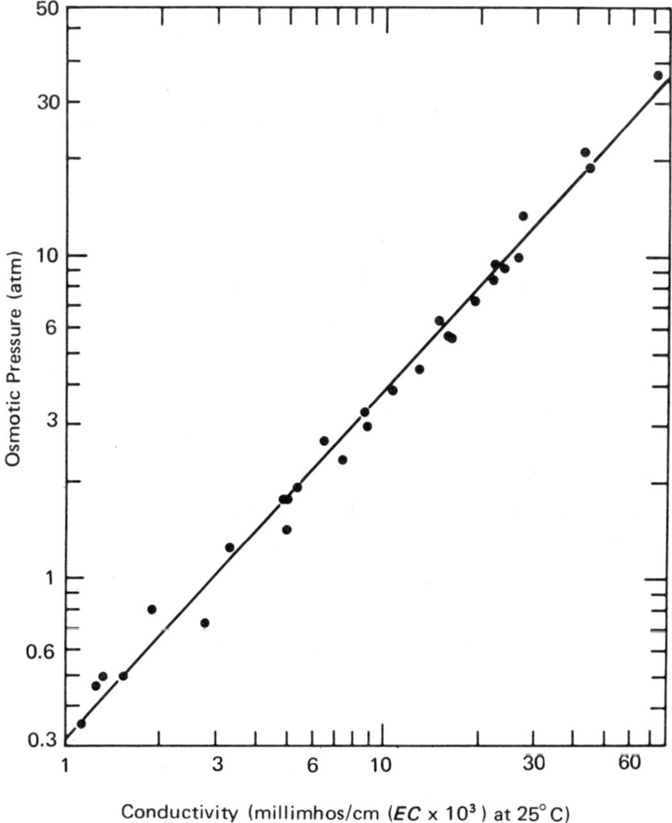

Figure II.3 Osmotic pressure of soil extracts as related to electrical conductivity.

In the range of *EC* that will permit plant growth,

$$OP = 0.36 \times EC \text{ (mmho/cm)} \qquad \text{(II.3)}$$

The electrical conductivity of aqueous salt solutions increases at the rate of approximately 2% per degree centigrade rise in temperature. The standard temperature for reporting electrical conductivity measurements is 25°C and is designated as EC_{25}. Temperature correction is usually made with the aid of correction tables (Richards, 1954) or a correction dial on the electronic measuring instrument.

Ionic composition

1. *Principal constituents*

The principal cations and anions in irrigation water are calcium, magnesium, sodium, bicarbonate, sulfate, chloride, and nitrate. If the pH exceeds 8.3, carbonate concentrations can become significant. Potassium concentrations are usually less than 1 meq/liter. Some general points to be considered in the analysis of waters follow:

i) Electroneutrality of the solution is always maintained. Thus, the total cation and anion concentrations, expressed in meq/liter, must be equal.

ii) In the past, sulfate concentrations were not often determined because of complex analytical procedures. This is no longer true. The turbidimetric method (Gilcreas *et al.*, 1965, page 291) is now routinely used at the U.S. Salinity Laboratory. The practise of estimating the sulfate concentration by difference between the sum of cations and anions should not be encouraged. Independent determination will assure the accuracy of the total chemical analysis since, for an accurate analysis, the sum of cations and anions will be equal.

iii) Calcium carbonates are sparingly soluble. The calcium concentration is very low in water of high pH (containing $CO_3{}^{2-}$), and is never high in the presence of a high concentration of bicarbonate. However, soil water extracts and water samples are often saturated with respect to calcium carbonates. Storage of samples at low temperature, rapid determination of Ca and $HCO_3{}^-$, and standard dilution, are recommended procedures to prevent a precipitation reaction from interfering with the analytical results.

iv) The electrical conductivity of the solution, in mmho/cm multiplied by 10, is approximately equal to the total cation or anion concentration in meq/liter.

v) The pH of irrigation water is not an accepted criterion of water quality because it tends to be buffered by the soil and most crops can tolerate a wide pH range.

2. *Minor constituents*

Some soluble trace elements may have an inhibitory effect on plant growth. Elements in this category include boron, lithium, selenium and some heavy metals. In areas suspected of contamination, water should be analyzed for these solutes. For detailed analytical procedures see Black (1965). Brief remarks concerning the more common trace elements are given below.

a) *Boron*

Boron is a constituent of practically all natural waters, in concentrations varying from minute traces to several parts per million. It is essential for plant growth but is exceedingly toxic at concentrations only slightly above optimum. The most widely applied criteria for boron in irrigation water are those published by Richards (1954). Later, Wilcox (1960) proposed a classification of crops according to their sensitivity to boron. This is shown in Table II.1.

Table II.1 Limits of boron in irrigation water for sensitive, semi-tolerant, and tolerant crop species based on toxicity symptoms observed on plants grown in sand culture

Sensitive 0.3–1 ppm boron	Semi-tolerant 1–2 ppm boron	Tolerant 2–4 ppm boron
Citrus	Lima bean	Carrot
Avocado	Sweet potato	Lettuce
Apricot	Bell pepper	Cabbage
Peach	Oat	Turnip
Cherry	Milo	Onion
Persimmon	Corn	Broad bean
Fig	Wheat	Alfalfa
Grape	Barley	Garden beet
Apple	Olive	Mangel
Pear	Field pea	Sugar beet
Plum	Radish	Palm
Navy bean	Tomato	Asparagus
Jerusalem artichoke	Cotton	
Walnut	Potato	
	Sunflower	

In each group the crops are arranged in descending order of tolerance within the range indicated.

It is known that an appreciable proportion of the boron added to soils as a component of irrigation water is fixed or sorbed by soil materials, with the balance remaining in the soil solution. Plants respond to the boron in soil solution independently of sorbed boron (Hatcher *et al.*, 1959). Hence, conditions affecting equilibria between sorbed and soluble boron are highly relevant to considerations of plant nutrition and quality criteria for irrigation water. Owing to adsorption, irrigation

Table II.2 Recommended maximum concentrations of trace elements in irrigation water[1]

Elements	For waters used continuously on all soil (mg/liter)	For use up to 20 yrs. on fine-textured soils at pH 6.0 to 8.5 (mg/liter)
Aluminum	5.0	20.0
Arsenic	0.10	2.0
Beryllium	0.10	0.50
Boron	0.75	2.0 – 10.0
Cadmium	0.010	0.050
Chromium	0.10	1.0
Cobalt	0.050	5.0
Copper	0.20	5.00
Fluorine	1.0	15.0
Iron	5.0	20.0
Lead	5.0	10.0
Lithium	2.5	2.5[2]
Manganese	0.20	10.0
Molybdenum	0.010	0.050[3]
Nickel	0.20	2.0
Selenium	0.020	0.020
Vanadium	0.10	1.0
Zinc	2.0	10.0

[1] These levels will not normally have an adverse effect on plants or soils. No data available for mercury, silver, tin, titanium, tungsten.
[2] Recommended maximum concentration for citrus is 0.75 mg/liter.
[3] Only for fine-textured acid soils, or acid soils with relatively high content of iron oxide.

water containing marginal levels of boron may not be immediately toxic. But after prolonged irrigation with such water, a new equilibrium will generally be established where the soluble boron levels will equal or exceed those of the irrigation water. It follows that, when this has occurred, plants more tolerant of boron must be grown, or the soil must be reclaimed. Prolonged use of water containing boron levels exceeding 3 ppm is not generally recommended.

b)   Other trace elements

In some areas, toxic levels of selenium occur in well waters. It has been reported that lithium in well water is toxic to citrus in some limited areas of California. Industrial pollution of natural water with copper, nickel, and other toxic heavy metals can also occur. However, the mere occurrence of toxic levels of trace elements in a water resource does not necessarily render the water unfit for agriculture. As water permeates the soil, the element may be precipitated, adsorbed, or fixed on the soil particles, and thus become inactive. Information on the chemistry of trace elements in the soils is limited, and the need for such information increases with the expanding use of effluent water for irrigation. Guidelines concerning the maximum permissible levels of various trace elements in irrigation water have recently been developed by a U.S. committee on water quality (Branson *et al.*, 1975), and are presented in Table II.2.

3.   Turbidity

Turbidity due to solid particles in suspension may restrict the use of water for irrigation. Solid particles may clog the water distribution systems, especially the nozzles of the sprinklers and drippers. Solid particles may also affect the permeability of the soil to air and water. Thus, special devices for the removal of sediments may be required.

Solution parameters derived from chemical composition

The absolute concentration values of the various cations in irrigation water are not sufficient for estimating potential hazards. An important consideration is the extent to which the exchangeable sodium percentage (*ESP*) of the soil will increase as a result of adsorption of sodium from the water. This increase depends on the ratio of soluble sodium to the divalent cations in solution. The higher the ratio, the higher the sodicity hazard. The sodicity hazard for a given cationic content increases when bicarbonate ions are present. The solution parameters related to soil properties are described in the following sections.

1. *Sodium adsorption ratio (SAR)*

Many equations have been proposed to describe the equilibrium distribution between the exchangeable and the soluble cations. Soils contain a mixture of different types of cation exchange materials and at least four cationic species. Consequently a rigorous theoretical description of ionic distribution is difficult, but reasonable success has been obtained with empirical equations. When soils are leached with salt solutions containing a mixture of monovalent and divalent cations until equilibrium is attained, a linear relation is obtained between the ratio of the concentrations of the exchangeable monovalent and divalent cations (*ESR* = exchangeable sodium ratio) and the ratio of the concentration (in meq/liter) of the monovalent cation to the square root of the concentration of the divalent cations in the solution (*SAR*).

$$ESR = a + k \, (SAR) \tag{II.4}$$

where a and k are empirical constants for a given soil type, and ESR is the exchangeable sodium ratio given by

$$ESR = \frac{[Na]}{[Ca + Mg]} = \frac{[Na]}{[CEC - Na]} \tag{II.5}$$

where [Na] and [Ca + Mg] are the concentrations of the exchangeable ions, and CEC is the cation-exchange capacity, all in meq/100 g of soil. SAR is the sodium adsorption ratio given by

$$SAR = \frac{(Na)}{\sqrt{\dfrac{(Ca + Mg)}{2}}} \tag{II.6}$$

where (Na) and (Ca + Mg) are the concentrations of the soluble ions in meq/liter.

Analysis of a large number of soil samples from Western USA (Richards, 1954) led to the empirical equation

$$ESR = -0.0126 + 0.01475 \, (SAR) \tag{II.7}$$

and is presented in Figure II.4. If a more accurate estimation of the exchangeable sodium ratio is required, the constants a and k of Eq. (II.4) should be determined experimentally for each soil.

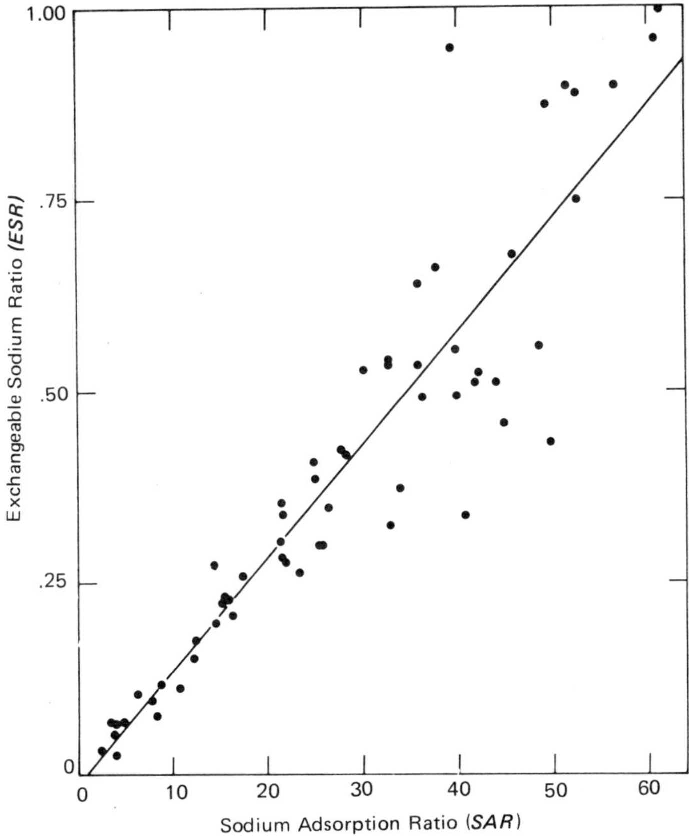

Figure II.4 Exchangeable sodium ratio (ESR) of soil samples from Western USA as related to sodium adsorption ratio (SAR) of soil extracts.

The k values for six different soils in Israel ranged between 0.0072 and 0.0169 (Fig. II.5).

A nomogram relating the ESP of the soils to the SAR of the irrigation water, based on Eq. (II.7), is given in Fig. II.6.

The SAR of irrigation water can be used as a measure of its sodicity hazards, providing it can be related to the resultant SAR of the equilibrated soil water (SAR_s). The concentration of the soil solution is increased by root water uptake and by evaporation. Depending on the frequency of irrigation, and the amount of water applied (see Chapter IV), the concentration of the water drained from the root zone may be two to ten times that of the irrigation water. However, the average concentration of the soil solution in the main root zone (about 50 cm depth) is not normally more than two or three times that of the irrigation water.

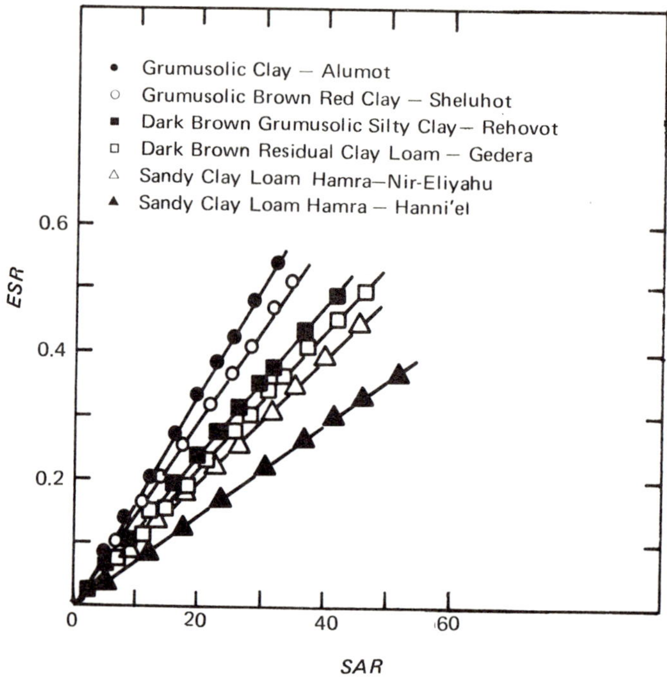

Figure II.5 *ESR* of six Israeli soils as related to the *SAR* of the soil saturation extract.

If we neglect the effect of salt precipitation or dissolution, mineral weathering, and the uptake of salts by plants, it is clear that when irrigation water enters the soil, it becomes more concentrated without change in relative composition, i.e., the soluble sodium percentage does not change. The *SAR* value, however, increases proportionally to the square root of the total concentration. Thus, if the total concentration is doubled, the *SAR* value will increase by a factor of 1.41. If the concentration is quadrupled, the *SAR* value will be doubled, etc. Since the soil *ESP* increases with increase in the *SAR*, it is obvious that the resultant *ESP* would be greater than the *SAR* of the irrigation water. This trend has been observed in lysimeters and field experiments (Rhoades, 1968; Richards, 1954).

14

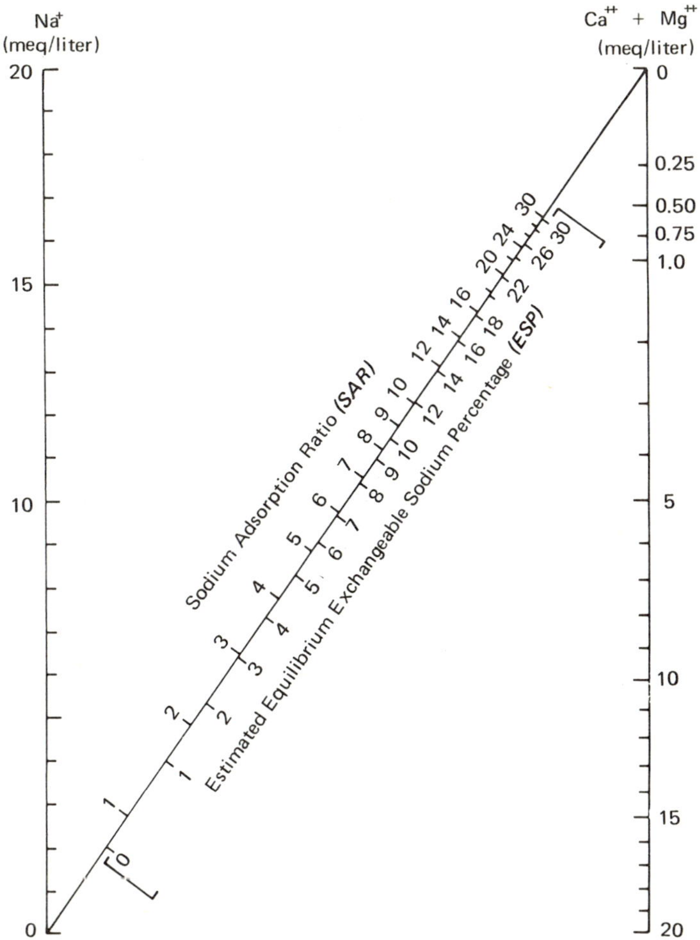

Figure II.6 Nomogram for determining the *SAR* value of irrigation water and for estimating the corresponding *ESP* value of a soil equilibrated with this water.

2. *Residual sodium carbonate (RSC)*

A major factor affecting the final *SAR* value of soil water is the change in calcium and magnesium concentration due to precipitation or dissolution of alkaline earth carbonates. In irrigation waters containing high concentrations of bicarbonate ions, there is a tendency for calcium and, to a lesser extent, magnesium to precipitate in the form of carbonate as the soil solution becomes more concentrated, thus leading to an increase in the *SAR* of the soil solution and consequently to an increase in the *ESP* of the soil. Although carbonate precipitation is common for many surface waters, its extent is often greater when well water is used. Underground water may be in equilibrium with a high partial pressure of CO_2 and a high bicarbonate concentration. On exposure to the atmosphere, the excess CO_2 is released, while calcium and magnesium carbonates precipitate. Eaton (1950) assumed that all calcium and magnesium would precipitate as carbonates and proposed the concept of residual sodium carbonate, *RSC*, for evaluating high carbonate water:

$$RSC = (CO_3^{2-} + HCO_3^-) - (Ca^{2+} + Mg^{2+}) \tag{II.8}$$

where the concentration of the ions is expressed in meq/liter. From the results of some limited leaching experiments, Eaton (1950) concluded that water with *RSC* higher than 2.5 is not suitable for irrigation purposes. He defined the range of *RSC* between 1.25 and 2.5 meq/liter as marginal and under 1.25 meq/liter as safe.

The concept of residual sodium carbonate has not proved to be very useful, mainly because of the assumption of quantitative precipitation. A better estimate of the precipitation tendency of carbonates is provided by the so-called "Saturation Index".

3. Saturation index (SI), pH_c^* and SAR

Bower *et al.* (1965, 1968) proposed the use of Langelier's saturation index to estimate carbonate precipitation from irrigation water as a function of the degree of $CaCO_3$ saturation of the soil solution. The index as applied to soil is:

$$SI = (8.4 - pH_c^*) \tag{II.9}$$

where 8.4 is the approximate pH of a nonsodic saline soil in equilibrium with $CaCO_3$, and is substituted for the pH of the water as originally proposed by Langelier. This substitution reflects the high buffering capacity of calcareous soils. The pH_c^* is defined by

$$pH_c^* = (pK_2 - pK_{sp}) + p(Ca + Mg) + p(CO_3 + HCO_3) \tag{II.10}$$

where $p(Ca + Mg)$ is the negative logarithm of the molar concentration of $Ca + Mg$, $p(CO_3 + HCO_3)$ is the negative logarithm of the equivalent concentration of CO_3 and HCO_3, and pK_2 and pK_{sp} are the negative logarithms of the second dissociation constant of H_2CO_3 and the solubility product of $CaCO_3$, respectively, both corrected for ionic strength (see Tables II.3 and II.4). The prediction equation for

Table II.3 Chemical parameters of some waters used for irrigation[1]

Origin	Ion concentration (meq/liter)											
	Na^+	K^+	Ca^{2+}	Mg^{2+}	Cl^-	SO_4^{2-}	HCO_3^-	EC mmho/cm	SAR	RSC	pH_c^*	SAR_{adj}
Grand River – USA	7.08	.19	2.0	.79	0.19	3.43	6.29	.94	6.0	3.50	7.32	10.5
Colorado River – USA	3.35	.22	6.95	3.63	1.03	9.31	3.73	1.27	1.5	− 6.8	7.03	3.5
Pecos River – USA	11.38	.08	15.98	9.07	12.13	22.39	3.11	3.26	3.2	−22.9	6.82	8.2
Nahal-Oz Well – Israel	41.30		1.3	3.7	35.20	1.5	9.6	4.60	26.1	4.6	7.06	61.0

[1] Equations (II.6), (II.8), (II.10), and (II.11) were used to calculate the SAR, RSC, pH_c^*, and SAR_{adj}, respectively. The data for Eq. (II.10) were taken from Table II.4.

exchangeable sodium, from the adjusted SAR values (SAR_{adj}), as given by Bower *et al.* (1968) is

$$ESP = SAR_{adj} = SAR_{iw} \, [1.0 + (8.4 - \text{pH}_c^*)] \qquad \text{(II.11)}$$

Bower *et al.* (1968) determined the effect of pH_c^* and leaching on the amount of $CaCO_3$ which precipitated. Fig. II.7 shows that as pH_c^* increases, precipitation of $CaCO_3$ falls off. In fact, at $\text{pH}_c^* > 8.4$, there is no precipitation of $CaCO_3$, and at $\text{pH}_c^* < 8.4$ there is a linear relationship between the amount of $CaCO_3$ precipitated and the value of the saturation index (SI).

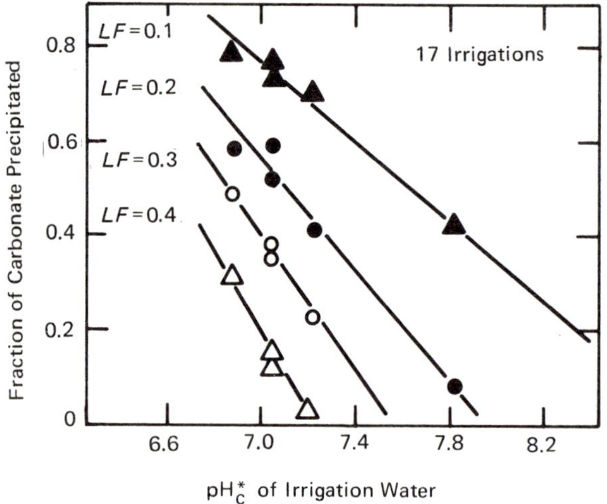

Figure II.7 Precipitation of $CaCO_3$ as related to pH_c^* of irrigation water and leaching fraction.

It is also evident from Fig. II.7 that as the leaching fraction (LF) (see Chapter V) increases, precipitation of $CaCO_3$ decreases. This effect was incorporated in the empirical equation

$$ESP = \frac{1}{\sqrt{LF}} \, [SAR_{iw} \, (1 + (8.4 - \text{pH}_c^*))] \qquad \text{(II.12)}$$

Specific examples of irrigation water quality

An example of the chemical composition of some waters used for irrigation in USA and Israel and their properties, as calculated from the equations given in Chapter II, are presented in Table II.3. Values for the different parameters for computing the values of pH_c^* are given in Table II.4.

Table II.4   Parameters for calculating pH_c^* from Eq. (II.10) for different values of the total ion concentration in the solution[1]

Concentration (meq/liter)	$(pk_2' - pk_c')$	$p(Ca + Mg)$	$p(CO_3 + HCO_3)$
0.1	—	4.30	4.00
0.5	2.11	3.60	3.30
1	2.13	3.30	3.00
2	2.16	3.00	2.70
4	2.20	2.70	2.40
6	2.23	2.52	2.22
8	2.25	2.40	2.10
10	2.27	2.30	2.00
15	2.32	2.12	1.82
20	2.35	2.00	1.70
25	2.38	1.90	1.60
30	2.40	1.82	1.52
35	2.42	1.76	1.46
40	2.44	1.70	1.40
50	2.47	1.60	1.30

[1] From Bower *et al.* (1965). The table relates $(pk_2' - pk_c')$ to total cation concentration, and $p(Ca + Mg)$ and pAlk to calcium and titratable base concentrations, respectively. The p value is the negative logarithm of ion concentration (moles/liter).

III. Soil Properties Affecting Water Quality

The texture of a soil (silty, loamy, sandy, etc.) is determined by the percentage distribution of the various particle sizes, designated as sand (>0.02 mm), silt ($0.02 - 0.002$ mm) and clay (<0.002 mm). It is evident that the smaller the particles, the higher is their specific surface area (surface area per gram of soil). In the case of montmorillonite clay, a dominant clay in arid and semiarid soils, the specific surface area may amount to as much as 800 m^2/g. Clay particles usually carry a net negative charge. This charge is neutralized by adsorbed cations, which can be exchanged for a stoichiometrically equivalent amount of other ions present in the solution. The reaction whereby a cation in solution replaces an adsorbed cation is called cation exchange.

Calcium and magnesium are the most prevalent cations in soil solutions and in exchange complexes of irrigated soils in arid regions. As a result of irrigation, sodium sometimes becomes the dominant cation in the soil solution, the salt concentration and the SAR tend to increase, and a part of the exchangeable calcium and magnesium is replaced by sodium.

The physical and mechanical properties of soils (such as dispersion of the particles, soil structure, stability of the aggregates) are very sensitive to the type of exchangeable ions. Divalent ions (mainly calcium) impart favorable physical properties to the soil, while adsorbed sodium causes clay dispersion and swelling. It is generally recognized that an exchangeable sodium percentage of 10 is sufficient to cause soil dispersion, reduction of soil permeability and impaired growth of some crop plants. On the other hand, high excess salt concentration prevents the dispersive effect of adsorbed sodium.

The purpose of the discussion which follows is to provide background knowledge on the physicochemical properties of soils for a better understanding of the effects of exchangeable sodium on soil properties.

Nature of clay minerals and origin of charge

Two structural units are involved in the atomic lattices of most clay minerals. One unit consists of two-dimensional arrays of silicon-oxygen tetrahedra. In each tetrahedron, a silicon atom is at equal distances from four oxygen atoms located at the four corners of the tetrahedron. Three of the four oxygen atoms are shared by three neighboring tetrahedra to form a tetrahedral sheet or a silica sheet (Fig. III.1a).

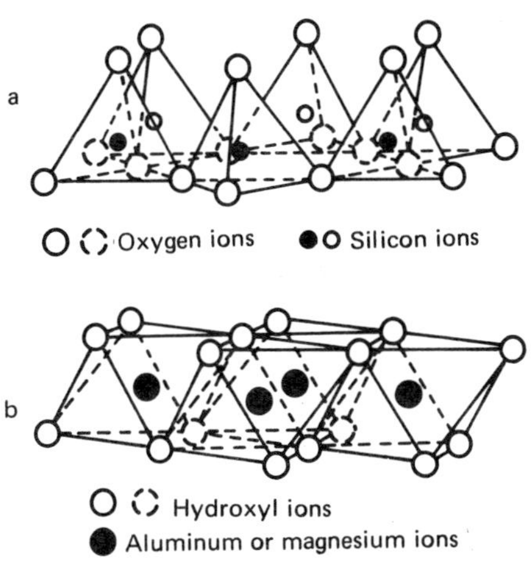

Figure III.1 Structural units in the atomic lattice of clay minerals: a) the silicon-oxygen tetrahedra; b) the aluminum-oxygen-hydroxyl octahedra.

The other unit consists of sheets of aluminum or iron- or magnesium-oxygen-hydroxyl octahedra. In each octahedron, the Al or Mg atoms are coordinated with six oxygen atoms or OH groups located around the central atom at the six corners of the octahedron. The sharing of oxygen atoms by neighboring octahedra results in an octahedral sheet (Fig. III.1b). When magnesium is present, all the octahedral positions are filled and the structure, with the formula $Mg_3(OH)_6$, is balanced. When aluminum is present, only two-thirds of the positions are filled to balance the structure, with the gibbsite formula $Al_2(OH)_6$.

The main clay minerals in the soil are classified according to the way in which these two sheets are stacked to form a unit layer. Grim (1968) classified clay minerals as follows:

1) Two-layer types. The unit layer is composed of one sheet of silica

tetrahedra and one sheet of alumina octahedra. The two sheets are held together by shared oxygen atoms. As shown for kaolinite in Fig. III.2, the surface of the layer on the alumina side consists of hydroxyl groups, while on the silica side it is composed of oxygen atoms. The layers show little tendency to separate, probably because of the hydrogen bonds between the adjacent oxygen and hydroxyl layers. Thus, kaolinite clays have a relatively low specific surface area.

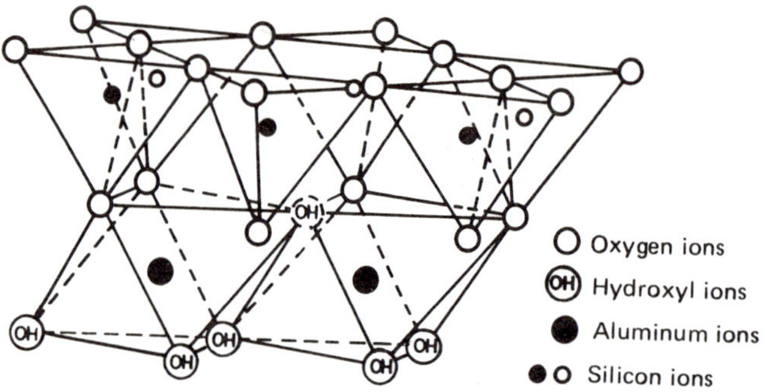

Figure III.2   The kaolinite crystal lattice.

2) Three-layer types. The unit layers are composed of two sheets of silica tetrahedra with an octahedral layer between. The three sheets are held together by shared oxygen atoms. Individual layers are stacked to form crystalline particles. Two groups of three-layer clays are recognized:

a) Expanding lattice or smectite group: e.g., montmorillonite, nontronite, hectorite. When montmorillonite clay is wetted, molecules of water penetrate between the unit layers, causing their separation, and the lattice expands. The extent of this expansion depends on the exchangeable ion.

b) Nonexpanding lattice or illite group. These clays are distinguished from the smectites mainly by the absence of an expanding interlayer.

Substitution of trivalent aluminum for quadrivalent silicon in the tetrahedral layer, and of magnesium or ferrous ions for aluminum in the octahedral layer (isomorphous substitution) results in a negative charge in the lattice. This charge is uniformly distributed within the lamellate clay particles. Charges also result from broken bonds at the edges of the

tetrahedral silica sheets. These surfaces are either positively or negatively charged, depending on the pH of the solution.

The particle charge is compensated by an equivalent amount of ions of
opposite sign, or counter ions, in the liquid immediately adjacent to the
particle. The total amount of counter ions expressed in milliequivalents
per 100 g of dry soil is known as the cation-exchange capacity (*CEC*) of
the soil. Chapman (in Black, 1965) summarized the experimental
methods and difficulties which may be encountered in analytically
determining the cation-exchange capacity.

Diffuse double layer

The diffuse double layer consists of the lattice charge and the
compensating counter ions. The counter ions are subject to two
opposing tendencies: (i) the cations are attracted electrostatically to the
negatively charged clay surface, and (ii) the cations tend to diffuse from
the surface of the particle, where their concentration is high, into the
bulk of the solution, where their concentration is low. The two
opposing tendencies result in decreasing counter ion concentration
from the clay surface to the bulk solution (Fig. III.3). A comparable
situation prevails in the earth's atmosphere where the air molecules are
subject to gravitational attraction and diffusion, with the result that the
concentration of the atmosphere gradually decreases away from the
earth's surface.

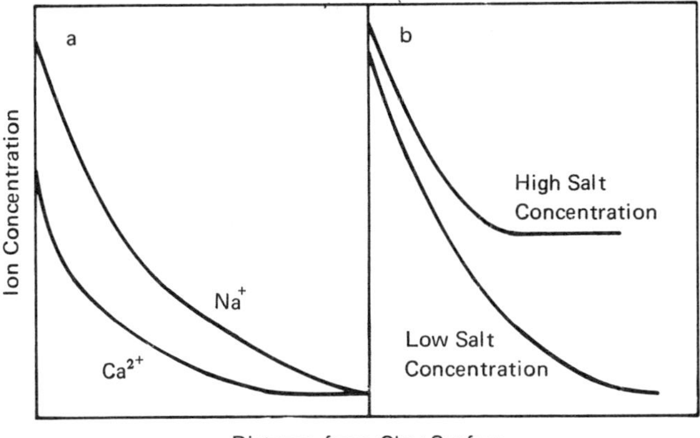

Figure III.3 Schematic representation of the ionic distribution
near a clay surface: a) mono- and divalent cations at the same salt
concentration; b) Na ions at two different salt concentrations.

Effect of counter-ion valency on the configuration of the double layer
Divalent ions are attracted to the surface with a force twice as large as that in the case of the monovalent ion. However, for the same electrolyte concentration in the bulk of the solution the tendency of the counter ions to diffuse away from the surface is the same. The net result is that the diffuse double layer in the divalent ion system is more compressed toward the surface (see Fig. III.3a).

Effect of salt concentration on the configuration of the diffuse double layer
The electrical attraction of the surface for counter ions (e.g., Na) is constant, irrespective of the bulk concentration. However, with increase in the electrolyte concentration in the bulk of the solution, the tendency of the counter ions to diffuse away from the surface is diminished. The net result is that the diffuse double layer is compressed towards the surface when the salt concentration in the bulk is increased (Fig. III.3b).

Swelling and dispersion of clay particles

1. *Homoionic Na and Ca systems*
When two clay particles approach each other, their diffuse counter ion atmospheres overlap as shown in Fig. III.4.

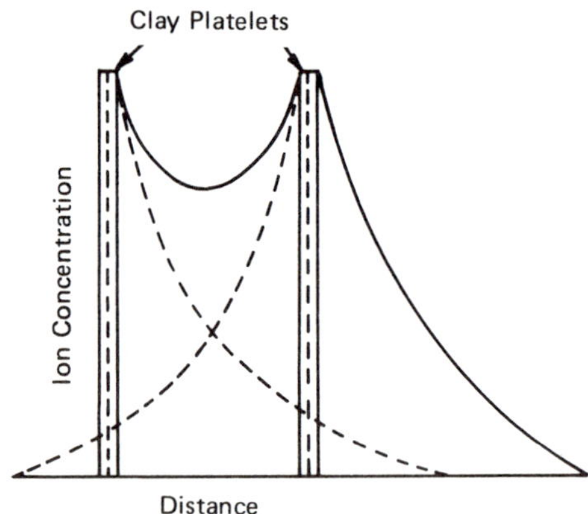

Figure III.4 Overlapping of two diffuse double layers (schematic): interrupted line — ionic concentration at a single platelet without overlapping; solid line — ionic concentration for overlapping platelets.

Work must be performed to overcome the electrical repulsion forces between the two positively charged ionic atmospheres. The electric double layer repulsion force, also called swelling pressure, can be calculated by means of the diffuse double layer theory. The greater the compression of the ionic atmosphere toward the clay surface, the smaller the overlap of the atmospheres for a given distance between the particles. Consequently, we can expect the repulsion force between the particles to decrease with increase in the salt concentration and in the valence of the adsorbed ions.

Because adsorbed sodium ions form a diffuse layer, high swelling pressures develop between Na-montmorillonite platelets, and single platelets tend to persist in dilute solutions.

On the other hand, the low swelling pressure between Ca-clay platelets prevents their dispersion, and they aggregate into packets or tactoids. Each tactoid consists of several clay platelets, with a 4.5Å thick film of water on each internal surface. The exchangeable Ca ions adsorbed on the internal surfaces of the tactoids do not form a diffuse layer.

2. *Mixed Na/Ca systems*

The soil solution in a saline soil usually contains a mixture of Na and Ca ions; it is therefore important to consider a mixed Na/Ca system. As indicated in the previous section, the platelets in sodium clays are separated when in equilibrium with dilute salt solution, whereas

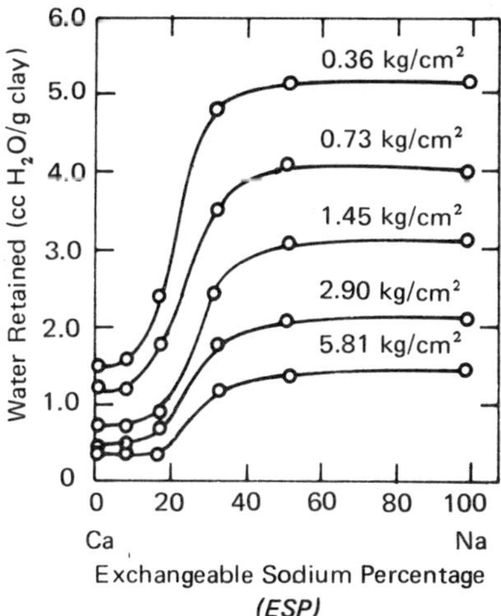

Figure III.5 Water retention by montmorillonite clay as a function of exchangeable sodium percentage (*ESP*) at various consolidation pressures.

calcium-saturated clays exist in tactoids, with only a fraction of the particle surfaces active in the swelling-shrinkage cycle. Hence, water retention increases with increasing exchangeable sodium percentage (Fig. III.5).

The retention curves for mixtures of mono- and divalent ions given in Fig. III.5 show clearly that a small addition of exchangeable sodium to a largely Ca-saturated clay has little effect on the amount of solution retained. Similarly, Shainberg and Otoh (1968) found that the introduction of a small amount of sodium into the exchange complex of Ca-tactoids had no effect on the size of the tactoids, while a larger amount of Na (over 20%) brought about a breakdown of the tactoids (Fig. III.6) and complete dispersion.

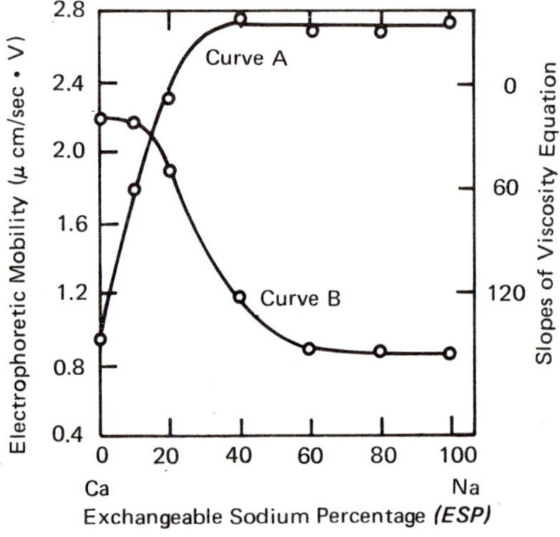

Figure III.6 The dependence of electrophoretic mobility in μ/sec/volt/cm (curve A) and the relative size (curve B) of montmorillonite particles on the exchangeable sodium percentage (*ESP*).

The electrophoretic mobility of clay particles, on the other hand, is strongly affected by small amounts of exchangeable sodium (Bar-On *et al.*, 1970; Fig. III.6). These phenomena indicate that the basic structure of the Ca-clay particles is not affected when 10% of the adsorbed calcium is replaced by sodium and that the sodium ions concentrate first on the outer surfaces of the tactoid imparting to it an electrophoretic mobility similar to that of Na-clay. Only further addition of sodium (above 20%) causes a breakdown of the tactoids and complete dispersion.

Hydraulic conductivity of salt-affected soils

The deleterious effect of adsorbed sodium on the physical properties of agricultural soils is most apparent in changes in their permeability. Soil permeability decreases with the square of the pore radius. Therefore, a small reduction in the size of the larger pores due to adsorbed sodium has a large effect on soil permeability. The reduction in porosity, especially of macropores, may result also in reduced aeration and gas exchange.

Gardner *et al.* (1959) reported that with a low percentage of sodium (<15) in the exchange complex, soil permeability increased less than twice as the solution concentration was increased from 2 to 100 meq/liter. At higher sodium contents permeability increased by several orders of magnitude with similar changes in salt concentration. Data obtained by Quirk and Schofield (1955) show similar results. The effect of solution concentration on soil permeability in soil with divalent cations was negligibly small.

The dependence of soil hydraulic conductivity on the total electrolyte concentration and the sodium adsorption ratio (*SAR*) of the percolating solution was assessed by McNeal and Coleman (1966) and McNeal (1968). The decrease in conductivity was particularly pronounced for soils with a high content of the swelling montmorillonitic clay, while soils high in nonswelling, kaolinite clay and sesquioxides were virtually insensitive to variations in solution composition.

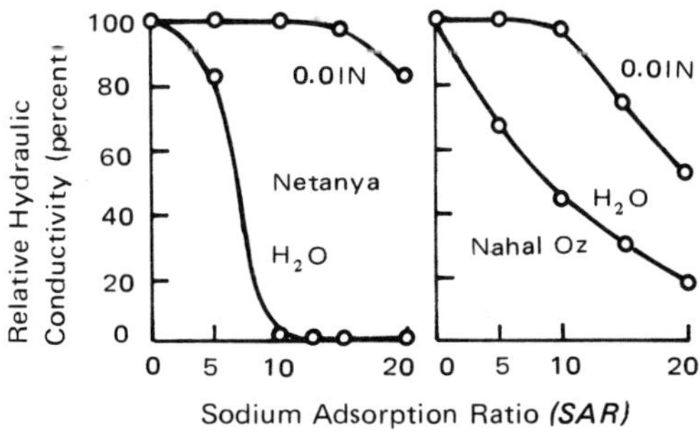

Figure III.7 Hydraulic conductivity of a sandy loam (Netanya) and a silty loam (Nahal Oz) soil as a function of the *SAR* and the concentration of the leaching solutions.

Double layer theory, confirmed by experiments, shows that an increased Na/Ca ratio in solution and decreased solution concentration result in both particle swelling and dispersion. Swelling results in reduced pore size while dispersion causes particle movement and pore blockage.

The effect of soil texture on the sensitivity of montmorillonitic soils to exchangeable Na was determined by Falhendler *et al.* (1974). They found that the reduction in the hydraulic conductivity was greater in a sandy loam soil (11.5% clay, 8 meq/100g *CEC*) than in a silty loam soil (16.4% clay, 10 meq/100 g *CEC*), despite the similarity in clay type and content. By analyzing the effluent from the two soils for clay content they showed that the difference between them was due to pore blockage as a result of greater clay particle dispersion and movement in the sandy loam than in the silty loam soil.

The difference between swelling and dispersion processes is quite important. Swelling is essentially a reversible process — reduction in permeability can be reversed by adding electrolytes or divalent ions to the system. Dispersion and particle movement on the other hand is essentially irreversible, causing the formation of impermeable clay pans.

Another demonstration of the importance of particle dispersion and movement as compared with swelling is given in Figure III.8. The reduction in permeability of a loam containing only 20% clay was much larger than that of a heavy clay soil containing 50% clay, when both were leached with distilled water.

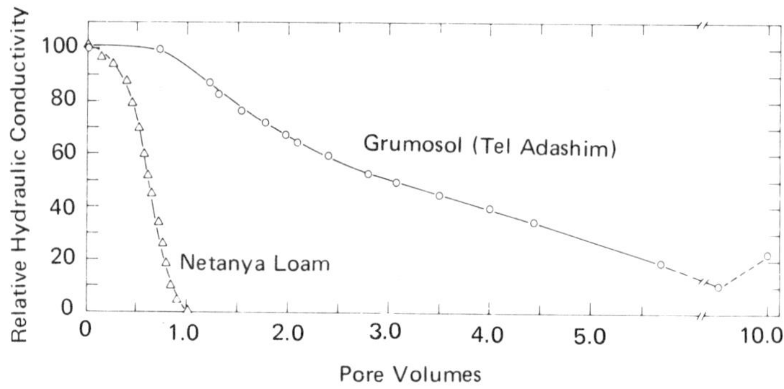

Figure III.8 Hydraulic conductivity of two montmorillonitic soils with *ESP* of 10, leached with distilled water.

The phenomenon demonstrated in Fig. III.8 has considerable practical importance. When high sodium water is used for irrigation, difficulties may not appear during the irrigation season. When winter rains leach the salts from the upper soil layer the soil disperses, the surface crusts and the permeability is drastically reduced.

Ion exchange equilibria

Ion exchange equations describe the distribution of cations between the adsorbed and the solution phase. The ionic equilibrium and the preference of the clay particles for one of the two counter ions can be described by the mass action law

$$\overline{Ca} + 2Na = 2\overline{Na} + Ca \tag{III.1}$$

where the bars indicate ions in the adsorbed phase. The selectivity coefficient is given by

$$K = \frac{[\overline{X}_{Na}]^2 \, (M_{Ca})}{[\overline{X}_{Ca}] \, (M_{Na})^2} \tag{III.2}$$

where the adsorbed ion concentration is expressed in units of ionic equivalent fractions, $\overline{X}$, and the concentration in the solution phase in moles per liter, M.

Levy and Hillel (1968) established a selectivity constant of 0.25 for a wide range of equivalent fractions of exchangeable sodium (between 0.1 and 0.7) in typical Israeli soils. It can be shown (Shainberg, 1973) that this constant is identical with the "Gapon" constant, 0.01475, recommended by the U.S. Salinity Laboratory in Eq. (II.7).

A characteristic of Eq. (II.6) is that, if constant composition of the surface phase is to be maintained, the *SAR* must be kept constant. However, by increasing the total concentration of the solution by a factor of two, the numerator is doubled whereas the denominator is only increased by the square root of two. Thus, the *SAR* of the solution is multiplied by a factor of $\sqrt{2}$, and more sodium is adsorbed on the surface. The data in Table III.1 show clearly that there is a preference in the exchange complex for the counter ion of higher valence and that this preference grows with dilution.

Table III.1 Effect of solution concentration on the equivalent fraction of cations adsorbed from solutions with a constant mole fraction (N) ($N_{Na} = N_{Ca} = 0.5$)

Solution concentration (mole/liter)	Equivalent fraction of adsorbed ion	
	X_{Na}	X_{Ca}
1.0	0.353	0.647
0.1	0.106	0.894
0.01	0.035	0.965
0.001	0.011	0.989

IV. Crop Growth and Salinity

Since the publication of Handbook 60 by the U.S. Salinity Laboratory staff (Richards, 1954), the salt tolerance of crops has been re-evaluated as a result of new data obtained under conditions of variable soil salinity (Bernstein and Francois, 1973b). It is now thought that most plants are able to withstand higher salinities of the drainage water than previously reported. Our discussion of interactions between crop growth and salinity is based on recent reviews by Bernstein (1974, 1975), Maas and Hoffman (1977) and Meiri and Shalhevet (in Yaron *et al.*, 1973).

Plant response to salinity

1. *Plant symptoms*

Herbaceous crops growing on saline soils may have barren spots, areas of stunted growth, and deep blue-green foliage. If the level of salinity is not sufficiently high to cause barren spots, the main characteristic of the crop may be marked variability in plant growth. However, with gravity irrigation, variable growth may also be caused by uneven water distribution due to inadequate levelling. Stunted growth and discoloration may also result from nutrient deficiency. Plants stunted because of low fertility are usually yellowish-green, while under conditions of high salinity they are characteristically bluish-green.

Leaf burn along the margin or at the tip, necrosis and defoliation are also indicative of salinity damage. These symptoms are usually associated with woody plants and result from excessive levels of sodium and chloride in the leaf tissues. They are more prevalent when the foliage is wetted and so, for example, leaf burn can often be seen in sprinkler-irrigated citrus groves.

Care must be exercised when plant symptoms are used to diagnose the extent or degree of soil salination. Overt symptoms appear mainly under severe salination; a reduction in yield may occur without any visible symptoms of salt injury. Soil salinity measurements, together with salt tolerance data, aid in the diagnosis of potential soil salinity problems.

2. Mechanism of salt injury

As water is removed from the soil by transpiration or by evaporation from the soil surface, the salt concentration of the soil solution in the root zone rises up to 2 to 5 times that in the irrigation water. Consequently, the osmotic potential of the soil water falls, while the concentration of specific ions potentially toxic to plant growth builds up. If growth depression is attributable to a decrease in osmotic potential, it is called an osmotic effect. If it is due to excessive concentration of specific ions, it is greater than that expected from osmotic effects alone, and is called a specific ion effect.

a) Osmotic effect

The osmotic effect has been demonstrated in several ways. For example, isoosmotic solutions of various salts are known to cause similar reduction in growth (Bernstein and Hayward, 1958); soil water matric and osmotic potentials have a similar and additive effect on plant growth (Wadleigh and Ayers, 1945); and water uptake and transpiration are reduced when the osmotic potential of the root media decreases (Meiri and Poljakoff-Mayber, 1970; Hoffman *et al.*, 1971).

The rate of water entry into plants depends on both the water potential gradient and the root resistance. When the osmotic potential of the root medium decreases without a corresponding decrease in the root water potential, the gradient for water flow from soil to root is reduced. For a constant resistance, the result is a reduced water uptake. This is one of the mechanisms of osmotic effects on plant growth. In many cases, the reduced soil osmotic potential induces a similar reduction in the leaf osmotic potential, keeping the gradient unchanged. This is known as osmotic adjustment (Bernstein, 1961), and results from an accumulation of inorganic or organic solutes in the plant sap. Bernstein (1975) considers osmotic adjustment as the primary mechanism affecting growth. Absorption of inorganic salts or synthesis of organic solutes to effect osmotic adjustment may cause reduced dry matter production (Gale *et al.*, 1967). Higher levels of respiration may also be needed to maintain the increased solute concentration. Excessive levels of salinity may also alter the hormone balance of plants (Bernstein, 1975), and it may damage plant cells and cytoplasmic organelles (Meiri and Shalhevet, 1973).

b)   Specific ion effect

Specific ion toxicities are commonly associated with woody species and result from excessive concentrations of Na or Cl ions (Bernstein, 1975). They rarely occur among herbaceous plants, but exceptions have been reported. Some annual crops which are not sensitive to chloride or sodium ions may develop symptoms of leaf burn when sprinkled with brackish waters containing 10-20 meq/liter of Na^+ or Ca^{2+}. Such damage has been observed in tomatoes (Gornat *et al.*, 1973), pepper (Goldberg and Shmueli, 1971; Bernstein and Francois, 1973a), and cotton (Busch and Turner, 1965).

Cation nutrition imbalances can also result from specific ion effects (Bernstein, 1975). Thus, calcium and magnesium deficiencies are liable to develop on nonsaline sodic soils due to the low concentration of these elements in the exchange and water soluble phases. Poor physical conditions, characteristic of nonsaline sodic soils, may also have an adverse effect on crop growth.

The mechanism of plant injury in the case of specific ion toxicities may involve injury to plant regulatory systems. Accumulation of chloride or sodium ions in the leaves of shrubs is believed to affect stomatal closure, causing excessive water loss and leaf injury symptoms similar to those of drought damage (Bernstein *et al.*, 1972). Many aspects of salinity effects on plant growth are not sufficiently understood, and continue to be the subject of intensive research.

3.   *Soil salinity and plant growth*

Soil salinity in the root zone varies with both depth and time. If the leaching fraction (LF) is greater than 0.3, the vertical distribution of electrical conductivity of soil water (EC_{sw}) is relatively uniform. For lower leaching fractions, the EC_{sw} at the bottom of the root zone may be 5 to 20 times that at the soil surface. Soil salinity also changes between irrigations because of evapotranspiration. The extent of the change depends on the amount and frequency of water application, as well as on the areal uniformity of water infiltration and on the rate of water uptake by plants. Thus, the greatest positional variations in EC_{sw} occur with furrow and trickle irrigation, due to salt accumulation between the irrigation furrows and between the tricklers. The most intensive root activity and the highest water uptake have been recorded in the zone of lowest soil salinity, beneath the furrow (Richards, 1954) or just below the drippers (Meiri and Shalhevet, 1973). Several studies indicate that plant response to variations in salinity is best related to the least saline zone.

Lunin and Gallatin (1965) and Bingham and Garber (1970) found that for corn and tomato a salinization of up to 2/3 of the root zone did not significantly affect the yield. Bernstein and Francois (1973b) suggested that plant response to salinity is related to a weighted mean salinity, based on water uptake. Since water uptake is highest at the upper part of the root zone, the weighted mean salinity is influenced more by the salinity in the upper portion of the root zone than by the higher salinity at the bottom. The EC_{sw} near the soil surface is approximately equal to that of the irrigation water. Consequently, the use of the new weighted mean salinity concept gives greater weight to the electrical conductivity of irrigation water, in terms of the effect on plant growth, than had been the case before.

Salt tolerance evaluations

Maas and Hoffman (1977) concluded from an extensive review of salt tolerance data that crop yield is not reduced until a threshold salinity

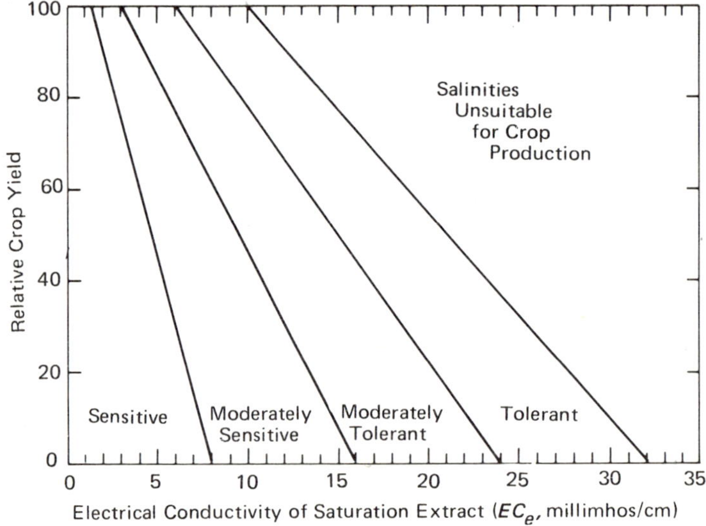

Figure IV.1 Relative crop yield as a function of the salinity of the soil saturation extract.

level is exceeded. Beyond the threshold level, the yield decreases linearly with rising salinity (Fig. IV.1). The salinity values at zero yield provide an estimate of maximum salinity that plants can tolerate, and are used to calculate leaching requirements. The division boundaries between different salt tolerance classes were chosen to conform with previously published terminology.

The salt tolerance values given in Table IV.1 have been calculated from the data of Mass and Hoffman (1977) and have been expressed according to the weighted average concept of Ayers and Branson (1975). Plant yields are largely determined by salinity levels in the upper part of the root zone. The upper portion of the root zone is defined as that which accounts for 2/3 of the water uptake. Through this depth increment the irrigation water is assumed to concentrate threefold.

Factors influencing salt tolerance

Plant response to salinity also depends on plant factors such as stage of growth, variety and rootstock, on soil factors such as fertility, water content and aeration, and on climate.

1. *Plant factors and tolerance tables*
Salinity effects in many crops depend mainly on the length of time plants are exposed to salinity regardless of the specific stage of growth (Shalhevet, 1970). Some other plants, however, are more sensitive at specific stages of growth. This is particularly true of cereals (Bernstein, 1974; Maas and Hoffman, 1977). Rice, barley, wheat, and corn are most sensitive during the early seedling stage and, in the case of rice, sensitivity may increase again during the flowering stage. Beet and sunflower are more sensitive to salinity during germination than in later stages of growth. However, localized high salt concentrations at seedling depth rather than a specific sensitivity may be the immediate cause of germination failure, which is common on saline soils (Bernstein, 1974). The tolerance of soybean may increase or decrease in the course of development, depending on the variety. Bernstein (1974) suggests a maximum permissible soil salinity of about 8 mmho/cm during the salt-sensitive stage of growth. The salt tolerance data in Table IV.1 were obtained from salinity treatments given after seedling establishment in nonsaline plots, so they do not necessarily apply to the germination and early seedling stages.

Table IV.1 Salt tolerance of agricultural crops.[1] Yield decrease to be expected due to salinity of irrigation water

Yield decrement:	0%			10%			25%			
Crop	EC_e[2]	EC_{iw}[3]	LR[4]	EC_e	EC_{iw}	LR	EC_e	EC_{iw}	LR	EC_d[5]
					Fruit Crops					
Date	4.0	2.7	4	6.8	4.5	7	11	7.3	12	64
Grapefruit	1.8	1.2	8	2.4	1.6	10	3	2.2	14	16
Orange	1.7	1.1	7	2.3	1.6	10	3	2.2	14	16
Apricot	1.6	1.1	9	2.0	1.3	12	3	1.8	15	12
Peach	1.7	1.1	9	2.2	1.5	11	3	1.9	15	13
Almond	1.5	1.0	7	2.0	1.4	10	3	1.9	14	14
Blackberry	1.5	1.0	8	2.0	1.3	11	3	1.8	15	12
Boysenberry	1.5	1.0	8	2.0	1.3	11	3	1.8	15	12
Grape	1.5	1.0	4	2.5	1.7	7	4	2.7	12	24
Plum	1.5	1.0	7	2.1	1.4	10	3	1.9	14	14
Strawberry	1.0	0.7	8	1.3	0.9	11	2	1.2	15	8

Vegetable Crops

Beet	4.0	2.7	9	5.1	3.4	11	7	4.5	15	30
Broccoli	2.8	1.9	7	3.9	2.6	10	6	3.7	14	27
Cucumber	2.5	1.7	8	3.3	2.2	11	4	2.9	15	20
Tomato	2.5	1.7	7	3.5	2.3	9	5	3.4	13	25
Spinach	2.0	1.3	4	3.3	2.2	7	5	3.5	12	30
Cabbage	1.8	1.2	5	2.8	1.9	8	4	2.9	12	24
Potato	1.7	1.1	6	2.5	1.7	8	4	2.5	13	20
Sweet corn	1.7	1.1	6	2.5	1.7	8	4	2.5	13	20
Pepper	1.5	1.0	6	2.2	1.5	9	3	2.2	13	17
Sweet potato	1.5	1.0	5	2.4	1.6	8	4	2.5	12	21
Lettuce	1.3	0.9	5	2.1	1.4	8	3	2.1	12	18
Onion	1.2	0.8	5	1.8	1.2	8	3	1.8	12	15
Radish	1.2	0.8	5	2.0	1.3	7	3	2.1	12	18
Carrot	1.0	0.7	4	1.7	1.1	7	3	1.9	11	16

Table IV.1 Continued

Yield decrement:	0%			10%			25%			
Crop	$EC_e{}^2$	$EC_{iw}{}^3$	LR^4	EC_e	EC_{iw}	LR	EC_e	EC_{iw}	LR	$EC_d{}^5$
Forage Crops										
Fairway wheatgrass	7.5	5.0	11	8.9	6.0	14	11	7.4	17	44
Tall wheatgrass	7.5	5.0	8	9.9	6.6	11	13	9.0	14	63
Bermudagrass	6.9	4.6	10	8.5	5.6	13	11	7.2	16	45
Barley	6.0	4.0	10	7.4	4.9	12	10	6.3	16	40
Perennial ryegrass	5.6	3.7	10	6.9	4.6	12	9	5.9	16	38
Birdsfoot trefoil	5.0	3.3	11	6.0	4.0	13	7	5.0	17	30
Hardinggrass	4.6	3.1	9	5.9	3.9	11	8	5.3	15	36
Tall fescue	3.9	2.6	6	5.8	3.9	9	9	5.7	13	46
Crested wheatgrass	3.5	2.3	4	6.0	4.0	7	10	6.5	11	57
Vetch	3.0	2.0	8	3.9	2.6	11	5	3.5	15	24
Sudangrass	2.8	1.9	4	5.1	3.4	7	9	5.7	11	52
Beardless wild rye	2.7	1.8	5	4.4	2.9	8	7	4.6	12	39
Big trefoil	2.3	1.5	10	2.8	1.9	13	4	2.4	16	15
Alfalfa	2.0	1.3	4	3.4	2.2	7	5	3.6	12	31
Lovegrass	2.0	1.3	5	3.2	2.1	8	5	3.3	12	28
Orchardgrass	1.5	1.0	3	3.1	2.1	6	5	3.7	11	35
Meadow foxtail	1.5	1.0	4	2.5	1.7	7	4	2.7	12	24
Alsike clover	1.5	1.0	5	2.3	1.6	8	4	2.4	12	20
Berseem clover	1.5	1.0	3	3.3	2.2	6	6	3.9	10	38

Field Crops

Barley	8.0	5.3	10	10.0	6.7	12	13	8.7	11	56
Cotton	7.7	5.1	10	9.6	6.4	12	13	8.3	16	54
Sugarbeet	7.0	4.7	10	8.7	5.8	12	11	7.5	16	48
Wheat	6.0	4.0	10	7.4	4.9	12	10	6.3	16	40
Soybean	5.0	3.3	17	5.5	3.7	18	6	4.2	21	13
Peanut	3.2	2.1	16	3.5	2.4	18	4	2.7	20	13
Rice (paddy)	3.0	2.0	9	3.8	2.6	11	5	3.4	15	23
Sesbania	2.3	1.5	5	3.7	2.5	8	6	3.9	12	33
Corn (grain)	1.7	1.1	6	2.5	1.7	8	4	2.5	13	20
Sugarcane	1.7	1.1	3	3.4	2.3	6	6	4.0	11	37
Broad bean	1.6	1.1	4	2.6	1.8	7	4	2.8	12	24
Flax	1.7	1.1	6	2.5	1.7	8	4	2.5	13	20
Cowpea	1.3	0.9	5	2.0	1.3	8	3	2.1	12	17
Bean (field)	1.0	0.7	5	1.5	1.0	8	2	1.5	12	12

[1] After Maas and Hoffman (1977).

[2] EC_e is the electrical conductivity of the saturation extract (mmho/cm at 25°C) in the root zone where about two-thirds of the water uptake occurs. For 0% yield reduction, EC_e is the threshold salinity at which yield is expected to begin to decline.

[3] EC_{iw} is the electrical conductivity of the irrigation water and was calculated from EC_e according to the expression $3\ EC_{iw} = 2\ EC_e$; the irrigation water is concentrated threefold in the root zone, which is equal to two times EC_e.

[4] Leaching requirement is calculated from Eq. (V.4).

[5] EC_d is the maximum electrical conductivity that can develop due to water uptake by the crop. At this EC, crop growth ceases.

Salinity generally reduces the growth rate and ultimate size of the plant, but this is often an unreliable guide for predicting fruit or seed production (Maas and Hoffman, 1977). For example, the effect of salinity on the vegetative growth of barley, wheat, cotton and some grasses is greater than on seed or fiber production, while a reversed relationship has been reported for rice and corn. Normally, salinity suppresses top growth more than root growth. However, in the case of root crops, salinity reduces the yield of root storage organs more than it does vegetative or root growth. Plant injuries resulting from specific salt effects often almost completely destroy yields, when vegetative growth has been only moderately reduced (Bernstein, 1974). Thus, a sulfate-induced calcium deficiency under saline conditions may cause total crop failure in tomatoes due to blossom-end rot, and in lettuce due to internal browning.

The type of rootstock of woody plants may affect the specific tolerance to chloride and sodium. Table IV.2 shows the chloride tolerance of a number of fruit crops. Salt-sensitive varieties may be injured when chloride in the saturation extract exceeds 5 meq/liter, whereas salt-tolerant ones may tolerate chloride concentrations of nearly 30 meq/liter. Leaf symptoms may develop when leaves contain 0.5 to 1% of chloride, or 0.25 to 0.5% of sodium calculated on the dry weight. Rootstocks can differ greatly in the absorption and transport of sodium and chloride (Bernstein *et al.*, 1969). Plants grown on root-stocks that absorb these ions slowly are more tolerant than those which accumulate them rapidly.

Rates of foliar absorption of chloride and sodium ions also vary, and again the higher the absorption rate the lower the salt tolerance. The rate of leaf absorption may exceed that of root absorption. In the case of deciduous fruit trees, chloride ion concentrations of 2-5 meq/liter in the irrigation water sprinkled on the leaves can cause leaf damage, whereas in the case of root absorption there is no harm with concentrations below 20 meq/liter (Table IV.2). In other crops, for example sugar cane, strawberries and avocadoes, foliar absorption is negligible.

Exchangeable sodium is also detrimental to crop growth. The effects of absorbed sodium are summarized in Table IV.3. The effect of sodium is generally indirect, as it affects plant nutrition or physical properties of the soil. The incidence of sodium-induced nutritional imbalance is dependent on the total salt concentration. In a nonsaline soil, with *EC* of about 1 mmho/cm and an exchangeable sodium percentage of 15,

Table IV.2 Maximum permissible chloride concentration in the saturation extract (Cl_e) of soil for fruit crop rootstocks and varieties if leaf injury is to be avoided (Bernstein, 1974)

Crop	Rootstock or variety	meq/liter
	Rootstock	
Citrus	Rangpur lime, Cleopatrà mandarin	25
(*Citrus* spp.)	Rough lemon, tangelo, sour orange	15
	Sweet orange, citrange	10
Stone fruits	Marianna	25
(*Prunus* spp.)	Lovell, Shalil	10
	Yunnan	7
Avocado	West Indian	8
(*Persea americana* Mill.)	Mexican	5
	Varieties	
Grape	Thompson Seedless, Perlette	25
(*Vitis* spp.)	Cardinal, Black Rose	10
Berries	Boysenberry	10
(*Rubus* spp.)	Olallie blackberry	10
	Indian Summer raspberry	5
Strawberry	Lassen	8
(*Fragaria* spp.)	Shasta	5

the concentrations of calcium and magnesium in the soil solution will be about 1 meq/liter, which is too low for some plants and may produce Ca deficiency symptoms. Sensitive crops, such as beans, are affected at *ESP* of about 10, and moderately tolerant to tolerant crops at *ESP* of about 25 and 50, respectively. Tolerance reflects the plant's ability to absorb adequate amounts of calcium and magnesium when the concentrations of these elements in the soil solution are low. In saline-sodic soils (*EC* > 4 mmho/cm) the concentrations of these ions are higher, and nutritional effects of sodium are usually absent. However, at *ESP* exceeding 15, poor soil structure may become the limiting factor, irrespective of the nutritional status of the plant.

Table IV.3 Tolerance of various crops to exchangeable sodium
percentage (*ESP*) under nonsaline conditions (Pearson,
1960)

Tolerance to *ESP* and range at which affected	Crop	Growth response under field conditions
Extremely sensitive (*ESP* = 2-10)	Deciduous fruits Nuts Citrus (*Citrus* spp.) Avocado (*Persea americana* Mill.)	Sodium toxicity symptoms even at low *ESP* values
Sensitive (*ESP* = 10-20)	Beans (*Phaseolus vulgaris* L.)	Stunted growth at low *ESP* values, even though the physical condition of the soil may be good
Moderately tolerant (*ESP* = 20-40)	Clover (*Trifolium* spp.) Oats (*Avena sativa* L.) Tall fescue (*Festuca arundinacea* Schreb.)	Stunted growth as a result of both nutritional factors and adverse soil conditions
Tolerant (*ESP* = 40-60)	Wheat (*Triticum aestivum* L.) Cotton (*Gossypium hirsutum* L.) Alfalfa (*Medicago sativa* L.) Barley (*Hordeum vulgare* L.) Tomatoes (*Lycopersicon esculentum* Mill.) Beets (*Beta vulgaris* L.)	Stunted growth, usually due to adverse physical condition of soil
Most tolerant (*ESP* = more than 60)	Crested and Fairway wheatgrass (*Agropyron* spp.) Tall wheatgrass (*Agropyron elongatum* (Host) Beau.) Rhodes grass (*Chloris gayana* Kunth)	Stunted growth, usually due to adverse physical condition of soil

2. *Soil factors*

Soil water content and frequency of irrigation can also affect the
tolerance of a crop to soil salinity. The drier the soil the lower are the
osmotic and the matric potentials of the soil solution. The inhibitory
effect of these potentials on plant growth tends to be additive.

Shortening the irrigation intervals minimizes the deleterious effect of both the osmotic and the matric potentials. The salt tolerance of pepper was enhanced when daily drip irrigation was used instead of less frequent furrow irrigation (Bernstein and Francois, 1973a).

The fertility level may also account for apparent differences in salt tolerance (Shalhevet, 1970; Bernstein, 1975; Maas and Hoffman, 1977). High apparent salt tolerance of plants grown on infertile soil may be due to the fact that fertility rather than salinity constitutes the limiting variable. Apparent decreases in salt tolerance as a result of excessive fertilization have also been reported (Maas and Hoffman, 1977). Bernstein (1975) concluded that increased nutrient levels do not significantly enhance salt tolerance, except when salinity induces a nutrient deficiency, or when poor soil physical conditions seriously impair root development, thus inducing nutrient deficiency.

3. *Climatic factors*
Climate can modify plant response to salinity (Maas and Hoffman, 1977). Salt tolerance is often reduced under hot, dry conditions, this being generally most pronounced in sensitive crops. Reduced salt tolerance in hot and dry weather has been reported for alfalfa, clover, bean, beet, cotton, squash and tomato (Bernstein, 1974). The onset of such weather may cause the sudden appearance of leaf burn in woody species (Bernstein, 1974).

Air pollution in the form of ozone is becoming increasingly prevalent around large urban areas. Soil salinity tends to counteract ozone damage in leafy vegetables and foliage crops. These crops may appear to be more salt tolerant in areas with air pollution than elsewhere (Maas and Hoffman, 1977).

V. Irrigation Management for Salt Control

Basic principles

The primary parameters that have to be considered to ensure effective irrigation management for salt control are the water requirement of the crop and the quality of the irrigation water. Correct irrigation should restore any soil water deficit, while avoiding the application of a wasteful and potentially harmful excess. An excess may be deliberately applied to control salt levels.

Although farming practices may vary from one irrigated area to another, the following principles of salt control have universal application (Richards, 1954). Plant growth is a function of the salinity and the matric potential of soil water. The salinity is controlled by leaching; matric potential is controlled by adequate and timely water application. Soluble salts contained in irrigation water are concentrated in the soil solution by soil evaporation and plant transpiration. Soluble salts are transported by water, and thus salinity control depends on the quality of the irrigation water and on the amount and direction of the water flow. Plant water uptake and surface evaporation may cause an upward flow, a process by which many soils become salinized, particularly when the water table is near the soil surface. The net water movement is downward and salts are leached from the root zone when more water is applied than is used during a crop season.

Over time, the amount of salt removed by leaching must suffice to prevent the build-up of salinity beyond the level the crop can tolerate. It was once thought that the amount removed had to equal the amount applied in the irrigation water.

However, recent lysimeter and model studies have shown that the amount of leached salt can be modified by chemical reactions, such as dissolution of soil minerals and salt precipitation. Mineral dissolution decreases and salt precipitation increases with reduction in the leaching fraction. Hence, under steady-state conditions, the amount of salt leached in drainage water may be greater than, equal to, or less than the amount of salt added from the irrigation water.

Chemical reactions also take place between the cations on the exchange complex of the soil and those in solution. Since the salt concentration in the soil solution increases with depth, there is a corresponding increase in the *SAR* of the soil solution in the same direction (see Chapter II). Hence, the *ESP* also increases in the lower part of the root zone and may attain levels detrimental to soil structure. Excessive levels of sodium in the soil solution may also be toxic to plant growth. Chemicals, such as gypsum, can be used to counteract these effects.

Salt balance and leaching requirement

1. *General equation*

Salt balance is determined by accounting for all the processes which contribute to the inflow, outflow and changes in salt in the profile (ΔS_g). This can be expressed mathematically as a mass conservation equation

$$\Delta S_g = D_r C_r + D_g C_g + D_i C_i + S_m - D_d C_d - S_p - S_c \qquad \text{(V.1)}$$

where D is the amount of water (in terms of depth of water spread uniformly over the soil surface); C is the salt concentration; and the subscripts r, i, d and g represent rain, irrigation, drainage, and upward movement from the ground water table, respectively; S_m is the amount of salt dissolved from soil minerals; S_p is the amount of salt that precipitates in the root zone; and S_c is the amount of salt removed in the harvested crop.

Assuming no appreciable contribution of salts from the dissolution of soil minerals, no loss of soluble salts by precipitation and crop removal, no rain and a water table depth sufficient to prevent upward movement of salt from ground water, Eq. (V.1) is reduced under steady-state to

$$D_i C_i - D_d C_d = 0 \qquad \text{(V.2)}$$

This equation states that, under the assumed conditions, in order to maintain salt balance the amount of salt added during irrigation must be equal to the amount drained. The importance of this concept is greater than its accuracy. It is implied that drainage or leaching must be ensured in irrigated agriculture, or else the salt concentration in the soil solution will ultimately reach toxic levels. Based on this concept, simple equations have been developed that relate the water needs of crops·and the salinity of irrigation water to the extent of leaching required.

2. *Leaching requirement*

Leaching fraction ($LF = D_d/D_i$) is the ratio between the amount of water drained below the root zone and the amount applied in irrigation. Under the conditions of insignificant rainfall and negligible chemical reaction, the EC of drainage water is controlled by the leaching fraction. It is evident from Eq. (V.2) that this ratio is also related to salt concentration in irrigation and drainage water:

$$LF = D_d/D_i = C_i/C_d = EC_i/EC_d \qquad\qquad (V.3)$$

It follows from the definition of LF and Eq. (V.3) that the values of EC_d may be equal for waters with different EC_i values, provided that the ratios EC_i/LF are equal. This was confirmed by results of controlled lysimeter experiments (Bernstein and Francois, 1973b) (Fig. V.1).

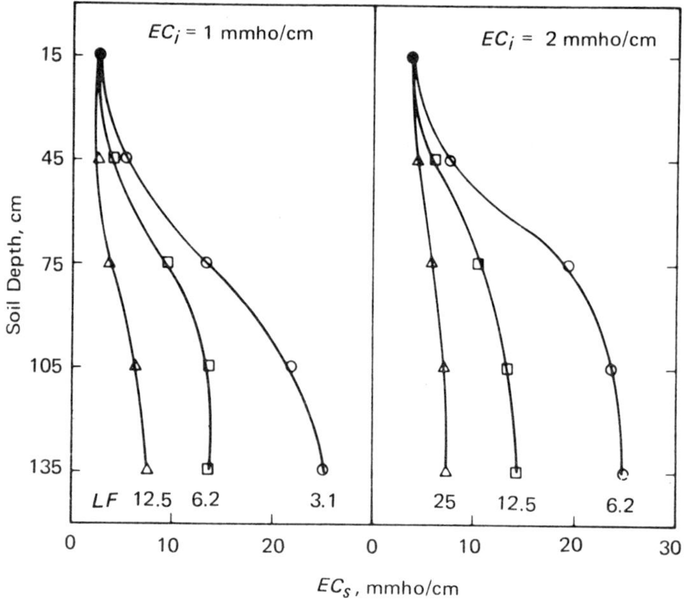

Figure **V.1** Salinity distribution in the soil profile as a function of the salinity of irrigation water and the leaching fraction.

Under steady-state conditions, the salinity tends to increase gradually from a level near the soil surface, controlled by the *EC* of the irrigation water, to one near the bottom of the root zone, which is determined mainly by the *LF*. Thus, by varying the leaching fraction it is possible to achieve some degree of control over the *EC* of the drainage water and the *EC* distribution in the root zone.

To estimate the minimum leaching requirement of an irrigated soil, one must first estimate the extra depth of irrigation water that has to be applied to maintain the average salt concentration of soil water below a level that would result in significant yield reduction. An estimate of this additional water requirement can be obtained from the following equation:

$$LR = EC_i/EC_d^* = D_d^*/D_i \tag{V.4}$$

Equation V.4 is similar to equation V.3 except that here the asterisk denotes required levels; EC_d^* denotes the maximum permissible salinity and D_d^* the minimum permissible amount of drainage water to maintain high yields.

While *LF* is the fraction of applied water that actually passes the lower boundary of the root zone, *LR* is an estimate of leaching required to maintain soil-water salinity within tolerable limits.

The evaluation of *LR* necessitates selection of appropriate EC_d^* values. These values vary according to the salt tolerance of the crop and can be estimated from the data supplied in Table IV.1.

Assuming uniform distribution of irrigation water, with no losses due to runoff, the amount of water needed for irrigation is equal to the amount required for evapotranspiration (D_{et}) and leaching or

$$D_i = D_{et} + D_d \tag{V.5}$$

By using Eq. (V.5) to eliminate D_d from Eq. (V.4), we obtain

$$D_i = \frac{D_{et}}{1\text{-}LR} \tag{V.6}$$

The irrigation requirement calculated from Eq. (V.6) is an estimate based on the salt tolerance and water requirement of the crop, and the salinity of the irrigation water. It does not take into account variability in water infiltration into the soil. Measured application efficiency seldom exceeds 85% and, with surface irrigation, it may be as low as 25%. Since the leaching requirements, as listed in Table IV.1 for waters with $EC < 3$ and for yield reductions of less than 10%, seldom exceed 15%, inefficiency in water application would seem to eliminate the need for more accurate estimates of LR. However, with increasing use of saline waters and of more efficient irrigation methods, the limitations of Eq. (V.6) acquire added importance.

To illustrate the use of the equations presented above, we may consider the following example. A cotton crop is to be produced with irrigation water having an EC of 4 mmho/cm. Using Table IV.1, EC_d is estimated to be 54 mmho/cm. Substitution of these values in Eq. (V.4) yields a value of 4/54. Thus, LR is approximately 0.08 or 8%. The LR value is then used to estimate the required depth of irrigation water and the minimum drainage requirement. If the consumptive use of water by cotton is 50 cm, then the minimum depth of drainage water D_d needed for salinity control (Eq. (V.4)) is 4 cm, and the desired depth of irrigation water is 54 cm.

The quality of irrigation water is affected by the amount of rain and its seasonal distribution. When rainfall is distributed throughout the growing season, a weighted average of rainfall and irrigation water quality is generally computed (Richards, 1954) as follows:

$$EC_{i+r} = \frac{EC_i D_i + EC_r D_r}{D_i + D_r} \qquad (V.7)$$

The value of EC_{i+r} is then substituted for EC_i in Eq. (V.4) to calculate the LR.

When the rainy season is short and concentrated, as in a Mediterranean type climate, rainfall may be sufficient to leach the salts out of the root zone. Under such conditions Eq. (V.7) is not applicable. Furthermore, soil type must be taken into account because of differences in the storage capacity of the soil (Shalhevet, 1973). For example, to displace once the pore volume at field capacity to a depth of 75 cm in a sandy soil (field capacity, weight basis = 12%) requires about 120 mm of effective rainfall, as against 320 mm for a clay soil (field capacity, weight basis = 32%). Assuming rain infiltration of 70%, the corresponding total rainfall for the two soils is 170 and 460 mm. Hence, if rainfall exceeds 300 mm during the rainy season, the leaching requirement for salinity control is of little account for the sandy soil but may be important for the clay soil. While rainfall is beneficial for salinity control, the reverse applies to potential sodicity hazards. This will be discussed below.

3.   *Salt removal by crops*

Salt removal by crops is mostly insufficient to maintain salt balance. It appears, however, that forage crops are capable of taking up an appreciable amount of salt, particularly when irrigation water is nonsaline. Data given by Pratt *et al.* (1967) for corn, and by Richards (1954) for alfalfa, indicate that the salt content of forages is about 7% of the dry weight. An annual alfalfa yield of 20 tons per hectare would thus remove 1.4 tons of salt. If we assume that 950 mm of irrigation water are applied per year, with an *EC* of 1 mmho/cm, the total salt application will be 6 tons (see Eq. (II.2)). The salt uptake by the crop would thus represent 20% of the amount applied.

4.   *Chemical reactions*

As indicated in Eq. (V.1), the salt balance also depends on chemical reactions involving slightly soluble salts, such as gypsum, lime, and silicate minerals. These reactions were studied in a three-year lysimeter experiment on soil cropped with alfalfa and irrigated with eight different irrigation waters at three leaching fractions (Rhoades *et al.*, 1974). Table V.1 records the observed chemical characteristics of the irrigation water and the resulting net and relative salt balances. If chemical reactions and salt uptake were insignificant, the amount of drained salt should equal the amount applied. The data show that the amount of salt in the drainage water may be greater or less than that applied in the irrigation water, indicating sources and sinks within the soil. For the least saline water the gain due to mineral dissolution was 300% at *LF* of 0.3. The extent of this phenomenon decreased with increasing salinity of the irrigation water and with decreasing leaching fraction.

When the *LF* was reduced to 0.1, in seven out of the eight cases there was a negative salt balance due to increased concentration of the soil solution, and precipitation of soil lime and gypsum. The composition of the drainage water, and hence also the amount of salt it contains, can be calculated from the solubility of soil lime and gypsum (Oster and Rhoades, 1975). The calculated and experimental salt burdens of drainage water for three irrigation waters are shown in Fig. V.2.

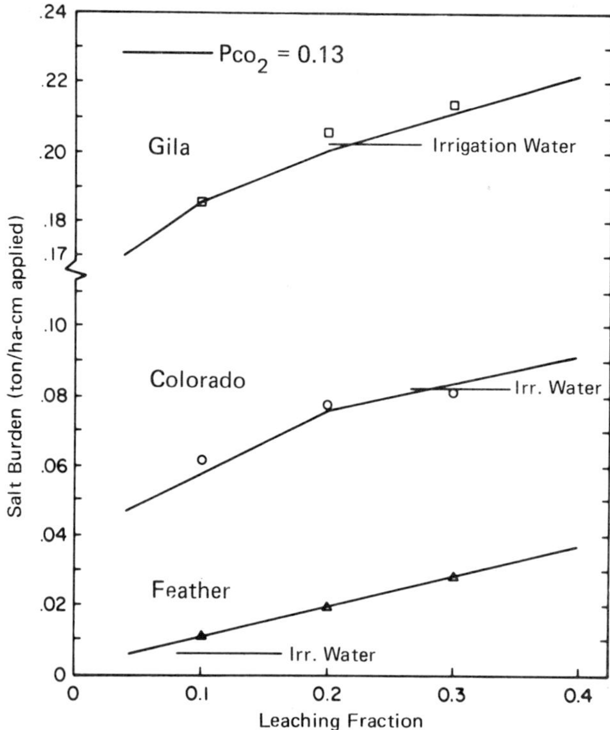

Figure V.2 The relationship between salt burden of the drainage water and leaching fraction for three irrigation waters.

The salt burden of the irrigation water is also indicated in Fig. V.2. The dependence of the salt burden on the *LF* suggests that irrigation management can be optimized to reduce the salt burden of irrigation return flows. These aspects were discussed by van Schilfgaarde *et al.* (1974).

Under average conditions, with moderate salinities and leaching fractions, the net effect of dissolution and precipitation is of little importance when estimating the leaching requirement. The use of Pecos River water with $EC = 3.3$ mmhos/cm is a case in point. Assuming no rainfall and chemical reactions, the calculated *LR* for beans and alfalfa are 0.27 and 0.10, respectively. Taking chemical reactions into account results in corrected *LR* of 0.18 and 0.04. The reduction is not significant, considering the present attainable field application efficiencies.

In the future, with increasing pressure on the currently available water

Table V.1 The electrical conductivities (EC), pH_c^*, salt burdens of applied water (D_iC_i) and the salt balances (the differences between the salt burdens of drainage water and those of applied water)[1]

Water	EC mmho/cm	Salt burden irr. water (tons/ha·m) D_iC_i	pH_c^*	D_dC_d (tons/ha·yr) $LF = 0.1$	$LF = 0.3$	Percent gain or loss $\dfrac{[D_dC_d - D_iC_i]}{D_iC_i} \times 100$ $LF = 0.1$	$LF = 0.3$
Feather	0.10	0.64	8.6	0.60	2.51	93	300
Grand	0.94	6.0	7.3	−1.66	0.40	−27	5
Missouri	0.91	5.8	7.3	−0.45	0.75	−8	10
Salt	1.56	10.0	7.5	−0.90	1.70	−9	13
Colorado	1.27	8.0	7.0	−2.06	−0.10	−25	−1
Sevier	2.03	13.0	6.9	−2.36	0.65	−18	4
Gila	3.14	20.0	7.1	−1.66	1.56	−8	6
Pecos	3.26	21.0	6.8	−8.28	−2.46	−40	−9

[1] From Rhoades *et al.* (1974).

[2] Based on a consumption of 900 mm/year

resources, irrigation efficiency will undoubtedly improve, resulting in a need for better control of salinity. One problem in obtaining good control of leaching is the difficulty of estimating crop water requirement with sufficient accuracy (Jensen, 1973; Doorenbos and Pruitt, 1975).

For example, a rather small 10% underestimation of D_{et} will result in no leaching at all when the needed leaching requirement is less than 10%. A long-term build-up in soil salinity would occur if only the calculated water were applied.

Where natural leaching by winter rains does not occur, a future irrigation control system may use soil salinity measurements to ensure adequate irrigation. Continual increases in root zone salinity will indicate insufficient water application, while salinities below the levels expected from the calculated LF will indicate over-irrigation (Oster and Rhoades, 1975). A rapid *in situ* measurement of soil salinity using a 4-probe electrical resistivity technique was proposed by Rhoades and Ingvalson (1971). Rhoades (1975) describes the technique and its usefulness for diagnostic, survey and management practices.

Sodicity hazard

The recommended equations for estimating the sodicity hazard are based on the pH_c^* concept discussed in Chapter II. The hazard is evaluated in terms of the estimated SAR for the upper (u) and lower (l) portions of the root zone, according to the equations

$$SAR_u = SAR_{iw} \left[1 + (8.4 - \mathrm{pH}_c^*) \right] \tag{V.8}$$

(See Eq. (II.11) and note that SAR is assumed to be equal to ESP) and

$$SAR_l = k \, SAR_u \tag{V.9}$$

The factor k (Rhoades, 1974) depends upon the leaching fraction, and is equal to 2.1, 1.4 and 1.0 for LF values of 0.1, 0.2, and 0.3, respectively. The effects of SAR on hydraulic conductivity depend on the salt concentration of the soil solution. Fig. V.3 shows combinations of salt concentration and ESP ($ESP \approx SAR$ for many soils) which resulted in a 25% reduction in hydraulic conductivity for several soils. Empirical and theoretical equations are available for estimating the reduction in hydraulic conductivity based on soil properties (McNeal, 1974).

However, these equations require the determination of the saturated hydraulic conductivity as a function of SAR and salt concentration under standard conditions. If a sodicity hazard is suspected, preventive action can be taken on the following lines:

1. Experimental determination of $EC - SAR$ threshold relations, of the type shown in Fig. V.3, on soils representative of the region to be irrigated;
2. Application of chemical amendments, such as gypsum, to increase the concentration of calcium in the soil solution.

Ayers and Branson (1975) suggest the following guidelines for assessing the permeability hazard of an irrigation water in terms of EC and calculated SAR_u of the irrigation water.

	EC (millimho/cm)	SAR_u
No problem	> 0.75	< 6.0
Increasing problems	0.75-0.3	6.9-9.0
Severe problems	< 0.2	> 9.0

It is possible to increase the calcium ion concentration of water with pH_c^* less than 8.4 by the addition of gypsum although this water is a saturated calcium bicarbonate solution. Some of the added calcium will precipitate as lime, thus reducing the bicarbonate ion concentration, and in turn permitting higher calcium levels in solution.

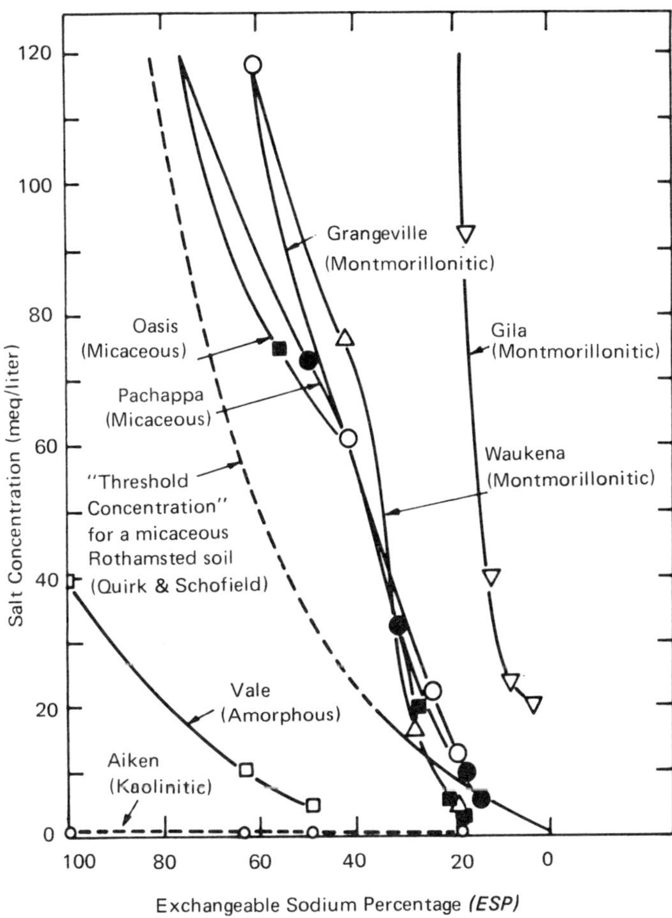

Figure V.3 Combinations of salt concentration and soil *ESP* required to produce a 25% reduction in hydraulic conductivity for selected soils.

Irrigation methods in relation to salinity control

The usability of poor quality water for irrigation is conditioned to a great extent by the irrigation method employed.

1. *Surface irrigation*

Efficient surface irrigation depends on an even distribution of water. Salinity control procedures include careful land leveling and controlled water application to ensure adequate and rapid watering of the border strip, basin or furrow with a minimum of runoff. Inadequate salinity control is mostly due to inherent variability of intake rates and unsatisfactory land leveling. High spots or areas of low intake rates receive inadequate water, whereas excess leaching occurs in areas of high soil permeability or at low spots. The application of additional water to assure adequate watering everywhere can result in an excessively high water table.

Artificial drainage is often required to control water table levels. The water table should be sufficiently low to prevent the rise of ground water into the root zone. Hence, the desirable water table level is dependent on the unsaturated conductivity of the soils. A relatively high water table (100 cm) can be tolerated in coarse-textured soils. A water table depth of about 180 cm is generally recommended for soils of medium texture. A lower table is required for perennial crops than for annuals (Shalhevet, 1973).

Inadequate watering and soil salinization may occur on fine-textured soils because of insufficient infiltration rates. This is particularly true of crops with high water requirements, such as alfalfa. If soil permeability is insufficient to meet the leaching requirement, a crop with a lower water requirement or higher salt tolerance should be grown. Another possibility is to grow crops with high water requirements in rotation with crops with lower water requirements. A more detailed discussion on the limitations imposed by soil properties on attainable leaching fractions is presented by Bernstein (1967) and by Rhoades (1974).

Furrow irrigation is commonly used for row crops. Since salts accumulate at the wetting front as water advances through the soil, salts tend to accumulate between the irrigation furrows (Fig. V.4).

Planting location on the ridge should be governed by the pattern of salt accumulation. The sloping bed system (Bernstein and Fireman, 1957) prevents salt accumulation at the location of seed placement. Rain during the growing season may cause damage by leaching salt into the root zone. The risk of damage can be greatly reduced by immediate irrigation. The salts which accumulate in the ridges during the cropping season will tend to become mixed throughout the surface soil when the field is prepared for the next season. The initial irrigation of the next crop will often provide adequate leaching for seed germination.

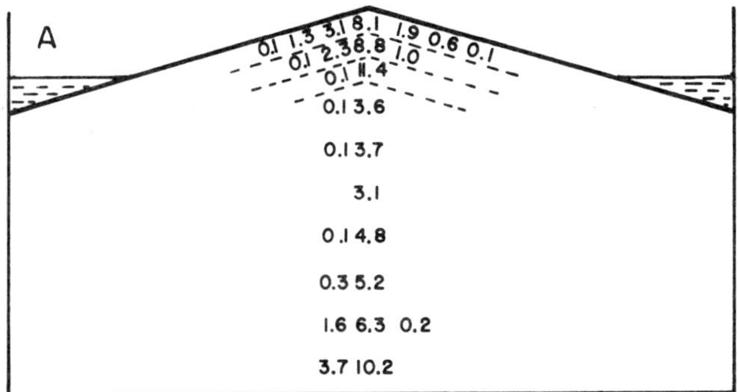

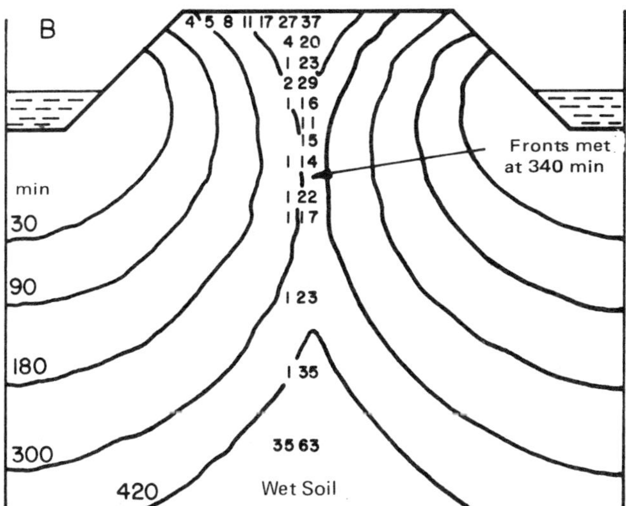

Figure V.4 Salt distribution under furrow irrigation (shapes of furrows A and B). The numbers denote the *EC* of the saturation extract.

2. *Sprinkler irrigation*

Sprinkler irrigation is becoming increasingly popular for a number of reasons. In countries with a limited water supply, water can be saved because with sprinkler irrigation higher efficiencies are easier to obtain than with flood irrigation. Land leveling is not necessary, and the need for artificial drainage is reduced — often natural drainage is sufficient.

Sprinkler irrigation provides an effective means of reducing the salt concentration at the soil surface; it is therefore increasingly used to start salt-sensitive row crops, such as lettuce. Furrow irrigation can then be used for the remainder of the growing season.

For irrigation of woody species, special low angle sprinkler heads can be used to minimize leaf burn due to wetting of the foliage.

3.   *Trickle irrigation*

In this method water is applied at low rates through emitters located near the base of the plant. Water deficits resulting from evapotranspiration can be replaced on a daily basis. Trickle irrigation has been successfully used with highly saline waters, $EC > 3$ mmho/cm. Recent data indicate that with this method higher levels of salinity can be tolerated by plants than with other irrigation methods (Bernstein and Francois, 1973a). It is believed that the increased tolerance is due to the steady supply of irrigation water which results in low salinity and high water content under the emitters. Under these conditions the uptake weighted mean salinity of the root zone is relatively constant and, for a leaching fraction of 0.1, it is about 2.5 times that of the irrigation water (Rawlins and Raats, 1975). Under flood irrigation, the uptake weighted mean salinity fluctuates from a minimum, which is about equal to the salinity of the irrigation water, to ten times this value prior to the next water application.

With trickle irrigation, zones of high salt concentration develop at the wetting front during the growing season and may have to be leached before a new crop is planted, except when row crops are planted in the old row location. In wheat sown after cotton, spotty germination of the wheat occurred although a presowing irrigation of 100 mm had been given (Yaron *et al.*, 1973).

Soil reclamation

1.   *Reclamation of sodic soils*

Reclamation of sodic soils involves replacement of exchangeable sodium by calcium. Calcium is added to the soil in the form of a soluble salt, or soil lime is solubilized by the addition of acid or acid-forming materials. The most common additive is gypsum ($CaSO_4 \cdot 2H_2O$), which is introduced into the soil or the irrigation water. If $CaCl_2$ is available at a low price, as in proximity to certain chemical industries, this salt can also be used. Acid or acid-forming additives (sulfuric acid, iron or aluminum sulfate or sulfur) can be applied if the soil contains lime. Reclamation of calcareous soils can also be achieved by growing rice under flood irrigation. Root and microbial activity raises the partial pressure of CO_2, which in turn increases the solubility of soil lime. However, continuous flood irrigation is only practicable in slowly permeable soils.

The following reactions illustrate how different additives react with the soil. When gypsum is added to the soil the reaction is

$$2NaX + CaSO_4 \rightleftharpoons CaX_2 + Na_2SO_4$$

where X represents an exchange site. Similar reactions occur with $CaCl_2$. When sulfuric acid is added to a calcareous soil, the reaction is

$$H_2SO_4 + CaCO_3 \rightleftharpoons CaSO_4 + CO_2 + H_2O$$

whereupon $CaSO_4$ reacts with exchangeable sodium. In the absence of lime the sulfuric acid and exchangeable sodium will react as follows:

$$H_2SO_4 + 2NaX \rightleftharpoons H_2X + Na_2SO_4$$

The resulting H-soil is not stable, and decomposes to form Al- or Fe-soil. When elemental sulfur is added to soil, it is slowly converted to sulfuric acid by microbiological oxidation according to two reactions,

$$2S + 3O_2 \rightleftharpoons 2SO_3$$

and

$$SO_3 + H_2O \rightleftharpoons H_2SO_4$$

and then the reaction with sulfuric acid proceeds as above.

The amount of additive needed can be estimated from the cation exchange capacity and the exchangeable sodium levels of the soil. As gypsum is most commonly used in reclamation, the following discussion is limited to this additive. When gypsum is incorporated into the soil and water is applied, the leachate is between one-third and one-half saturated with gypsum (Quirk and Schofield, 1955; Chaudhry and Warkentin, 1968). Hence, 1 hectare-m of irrigation water will dissolve between 8.6 and 13.2 metric tons of gypsum. It takes 15.6 tons of gypsum per hectare to replace 1 meq/100 g of exchangeable sodium in the soil to a depth of 1 m, assuming 75% efficiency. If the leaching solution is one-third saturated, 1.8 m of water would be needed to achieve such reclamation. The actual amount of excess exchangeable sodium depends on the final exchangeable sodium that is needed to assure a satisfactory soil hydraulic conductivity. This can be determined from hydraulic conductivity relationships for a given soil. Under most conditions, *ESP*'s of 5 and 15 at soil depths of 0.2 and 1 m are adequate.

Reclamation in stages is generally recommended for sodic soils. During the first stage only the top soil is reclaimed. Further reclamation may be done with a crop growing in the field.

Another method of reclamation consists in leaching the soil with successive dilutions of highly saline water such as sea water (Reeve and Bower, 1960). In the early phase, the high salinity of the water prevents dispersion and induces flocculation of the soil colloids, and the calcium content provides a source of calcium for exchange with sodium. On dilution, the *SAR* of the water is reduced and, provided the divalent ion concentration of the saline water is at least 30% of the total cation concentration, satisfactory *SAR* levels are ensured in the final dilution. If the dilution factor, d, is expressed as

$$d = \frac{V_w}{V_s} + 1 \qquad\qquad (V.10)$$

where V_w is the volume of salt-free water used for dilution, and V_s the volume of saline water, then

$$SAR_{dil} = SAR_s/d \qquad\qquad (V.11)$$

When sufficient water is applied at each step of dilution, then, to ensure exchange equilibrium to the depth desired, the total equivalent surface depth of water for reclamation should be about nine times the depth of soil to be reclaimed (Reeve and Doering, 1966). This amount can be reduced if complete equilibration is not obtained for each dilution step, and if the water is saturated with gypsum (Muhammed *et al.*, 1969).

2.   *Reclamation of saline soils*
Saline soils are normally reclaimed by flooding. The depth of soil leached is approximately equal to the depth of water infiltrated during the leaching period. The displacement of one pore volume of soil water

reduces the salinity by about one-half, and the displacement of 1.5 to 2.0 pore volumes reduces the salinity by about 80%. Since the relative pore volume of soils is about 50%, a displacement of 2 pore volumes is approximately equal to an equivalent depth of water per unit depth of soil. This is illustrated by field data obtained by Reeve *et al.* (1955) for a highly saline ($EC_e > 40$ mmho/cm) silty clay loam (Fig. V.5).

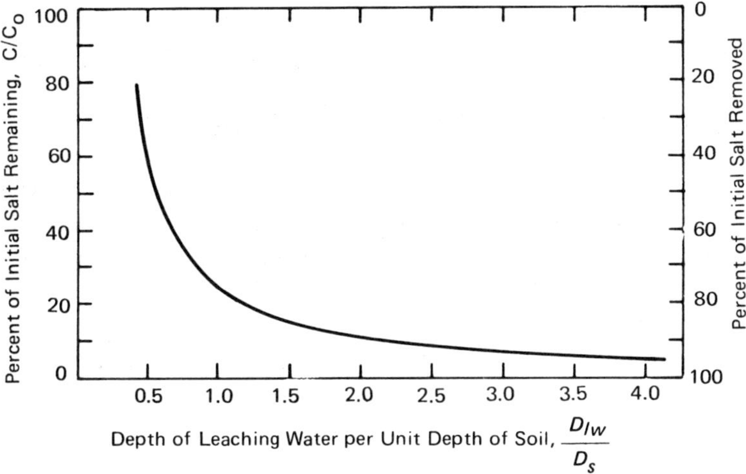

Figure V.5 Depth of water per unit depth of soil required to leach a highly saline soil.

Leaching by intermittent ponding or sprinkling uses less water than continuous ponding. Miller *et al.* (1965) studied the efficiency of leaching for different flood irrigation techniques on a clay loam soil with a saturated hydraulic conductivity of about 30 cm/day. Intermittent ponding of 5 and 15 cm of water required, respectively, 33% and 20% less water than continuous ponding. A reduction of 50% and 30% for intermittent ponding and sprinkling on a silty clay soil was reported by Oster *et al.* (1972). The time required to reduce salinity by 50% was the same for all treatments on the silty clay soil, whereas reclamation by continuous ponding proved to be more expeditious on the more permeable clay loam soil. Whereas continuous ponding necessitates extensive soil borders, intermittent ponding can often be performed with existing border strips, thus reducing field preparation costs.

The superior performance of intermittent ponding and sprinkling is attributable to the fact that salt is displaced by water more effectively when the soil is partially desaturated. The rate of salt movement depends on the velocity of the water flow (Bresler, 1973). This velocity is conditioned by the porous structure of the soil; the flow is faster in the large pores than in small ones, so that the wetter the soil the greater the relative amount of water moving in large pores. Salt displacement by water movement in large pores is less efficient than in small pores; hence, the wetter the soil the less efficient the leaching process.

When natural drainage is limiting, leaching efficiency of flood and intermittent irrigation is also dependent on the artificial drainage system. Since the hydraulic gradient is greater over tile drains, more water enters the soil over the drains than between them. Application of 30 cm of water in a field with a tile line spacing of 27 m removed 74% of salt from 60 cm depth near the tile lines, while only 20% was removed from the same depth between the tile lines (Talsma, 1967). In another field, with a tile spacing of 20 m and water application of 22 cm, the percentages of salt removed were 73% and 55%, respectively. Thus, leaching improves as the spacing decreases. If leaching between tile lines becomes limiting, then leaching the areas between the drains may have to be considered.

VI. References

Ayers, R.S. and Branson, R.L. (1975) Guidelines for interpretation of water quality for agriculture. University of California. 13 pp

Bar-On, P., Shainberg, I. and Michaeli, I. (1970) The electrophoretic mobility of Na/Ca montmorillonite particles. *J. Colloid and Interface Sci.* **33**: 471-472

Bernstein, L. (1961) Osmotic adjustment of plants to saline media. I. Steady-state. *Amer. J. Bot.* **48**: 909-918

Bernstein, L. (1967) Quantitative assessment of irrigation water quality. *Amer. Soc. Testing Mater., Spec. Tech. Pub.* **416**: 51-65

Bernstein, L. (1974) Crop growth and salinity. Chapter 3 in: van Schilfgaarde, Jan (ed.), Drainage for Agriculture. *Agronomy 17, Amer. Soc. Agron.* Madison, Wisconsin

Bernstein, L. (1975) Effects of salinity and sodicity on plant growth. *Am. Rev. of Phytopathology* **13**: 295-311

Bernstein, L., Ehlig, C.F. and Clark, R.A. (1969) Effect of grape rootstocks on chloride accumulation in leaves. *J. Am. Soc. Hortic. Sci.* **94**: 584-590

Bernstein, L. and Fireman, M. (1957) Laboratory studies on salt distribution in furrow irrigated soil with special reference to pre-emergence period. *Soil Sci.* **83**: 249-263

Bernstein, L. and Hayward, H.E. (1958) Physiology of salt tolerance. *Ann. Rev. Plant Physiol.* **9**: 25-46

Bernstein, L., Francois, L.E. and Clark, R.A. (1972) Salt tolerance of ornamental shrubs and ground covers. *J. Am. Soc. Hortic. Sci.* **97**: 550-556

Bernstein, L. and Francois, L.E. (1973a) Comparison of drip, furrow and sprinkler irrigation. *Soil Sci.* **115**: 73-86

Bernstein, L. and Francois, L.E. (1973b) Leaching requirement studies: Sensitivity of alfalfa to salinity of irrigation and drainage waters. *Soil Sci. Soc. Amer. Proc.* **37**: 931-943

Bingham, F.T. and Garber, M.J. (1970) Zonal salinization of the root system with NaCl and boron in relation to growth and water uptake of corn plants. *Soil Sci. Soc. Am. Proc.* **34**: 122-126

Black, C.A. (ed.) (1965) Methods of soil analysis. *Agronomy 9, Amer. Soc. Agron.* Madison, Wisconsin

Bower, C.A., Ogata, G. and Tucker, J.M. (1968) Sodium hazard of irrigation waters was influenced by leaching fraction and by precipitation or solution of calcium carbonate. *Soil Sci.* **106**: 29-34

Bower, C.A., Wilcox, L.V., Akin, G.W. and Keyes, Mary G. (1965) An index of the tendency of $CaCO_3$ to precipitate from irrigation waters. *Soil Sci. Soc. Amer. Proc.* **29**: 91-92

Branson, R.L., Pratt, P.F., Rhoades, J.D. and Oster, J.D. (1975) Water quality in irrigated watersheds. *J. Environ. Quality* **4**: 33-40

Bresler, E. (1973) Solute movement in soils. Chapter IV.2 in Yaron, B., Danfors, E. and Vaadia, Y. (eds.), Arid Zone Irrigation. Springer Verlag, Berlin. 433 pp

Busch, C.D. and Turner, F., Jr. (1965) Sprinkling cotton with saline water. *Prog. Agric. Ariz.* **17** (4): 27-28

Chaudhry, G.H. and Warkentin, B.P. (1968) Studies on exchange of sodium from soils by leaching with calcium sulfate. *Soil Sci.* **105**: 190-197

Doorenbos, J. and Pruitt, W.O. (1975) Crop-water requirements. FAO Irrigation and Drainage Paper No. 24, Rome, Italy, 179 pp.

Eaton, F.M. (1950) Significance of carbonates in irrigation waters. *Soil Sci.* **69**: 123-133

Falhendler, Rachel, Shainberg, I. and Frenkel, H. (1974) Dispersion and the hydraulic conductivity of soils in mixed solutions. *Proc. 10th Intern. Cong. Soil Sci.* Moscow, USSR, Vol. II, 103-111

Gale, J., Kohl, H.C. and Hagan, R.M. (1967) Changes in the water balance and photosynthesis of onion, bean and cotton plants under saline conditions. *Physiol. Plant.* **20**: 408-420

Gardner, W.R., Maybaugh, M.S., Goertzen, J.O. and Bower, C.A. (1959) Effect of electrolyte concentration and exchangeable sodium percentage on diffusivity of water in soils. *Soil Sci.* **88**: 270-274

Gilcreas, F.W., Taras, M.J., Ingals, R.S. and Orland, H.P. (eds.) (1965) Standard methods for the examination of water and wastewater (12th ed.). American Public Health Association, Inc. New York

Goldberg, D. and Shmueli, E. (1971) Sprinkle and trickle irrigation of green pepper in the arid zone. *Hort. Sci.* **6**(6): 559-562

Gornat, B., Goldberg, D., Rimon, D. and Ben-Asher, J. (1973) The physiological effect of water quality and method of application on tomato, cucumber and pepper. *J. Am. Soc. Hort. Sci.* **98**: 202-205

Grim, R.E. (1968) Clay Mineralogy (2nd ed.), McGraw-Hill, New York

Hatcher, J.T., Blair, G.Y. and Bower, C.A. (1959) Response of beans to dissolved and adsorbed boron. *Soil Sci.* **88**: 98-100

Hoffman, G.J., Rawlins, S.L., Garber, M.S. and Callen, E.M. (1971) Water relations and growth of cotton as influenced by salinity and relative humidity. *Agron. J.* **63**: 822-826

Jensen, M.F. (1973) Consumptive use of water and irrigation water requirements. Techn. Com. on Irrigation Water Requirements. Irrigation and Drainage Div. ASCE, 215 pp

Langelier, W.F. (1936) The analytical control of anti-corrosion water treatment. *J. Amer. Waterworks Assn.* **28**: 1500-1521

Levy, R. and Hillel, D. (1968) Thermodynamics equilibrium constants of Na/Ca exchange in some Israeli soils. *Soil Sci.* **106**: 393-398

Lunin, J. and Gallatin, M.H. (1965) Zonal salinization of the root system in relation to plant growth. *Soil Sci. Am. Proc.* **29**: 608-612

Maas, E.V. and Hoffman, G.J. (1977) Crop salt tolerance — Current assessment. *J. Irrig. and Drainage Div., ASCE* **103 (IR2)**: 115-134

McNeal, B.L. (1968) Prediction of the effect of mixed-salt solutions on soil hydraulic conductivity. *Soil Sci. Soc. Amer. Proc.* **31**: 190-193

McNeal, B.L. (1974) Soil salts and their effects on water movement. Chapter 15 in: van Schilfgaarde, Jan (ed.), Drainage for Agriculture. *Agronomy 17, Amer. Soc. Agron.,* Madison, Wisconsin

McNeal, B.L. and Coleman, N.T. (1966) Effect of solution composition on soil hydraulic conductivity. *Soil Sci. Soc. Amer. Proc.* **30**: 308-312

Meiri, A. and Poljakoff-Mayber, A. (1970) Effect of various salinity regimes on growth, leaf expansion and transpiration rate of bean plants. *Soil Sci.* **109**: 26-34

Meiri, A. and Shalhevet, J. (1973) Irrigation with saline water. Chapter VI.2 in: Yaron, B., Danfors, E. and Vaadia, Y. (eds.), Arid Zone Irrigation. Springer Verlag, Berlin, 433 pp

Miller, R.J., Nielson, D.R. and Biggar, J.W. (1965) Chloride displacement in Panoche clay loam in relation to water movement and distribution. *J. Water Resour. Res.* **1**: 63-73

Muhammed, Shah, McNeal, B.L., Bower, C.A. and Pratt, P.F. (1969) Modification of the high-salt-water method for reclaiming sodic soils. *Soil Sci.* **108**: 249-256

Oster, J.D. and Rhoades, J.D. (1975) Calculated drainage water composition and salt burdens resulting from irrigation with river waters in the western United States. *J. Environ. Qual.* **4**: 73-79

Oster, J.D., Willardson, L.S. and Hoffman, G.J. (1972) Sprinkling and ponding techniques for reclaiming saline soils. *Trans. of ASAE* **15**: 1115-1117

Pearson, G.A. (1960) Tolerance of crops to exchangeable sodium. *USDA Inf. Bull.* **216**: 4

Poljakoff-Mayber, A. and Gale, J. (1975) Plants in Saline Environments. Springer Verlag, Berlin, New York, 213 pp

Pratt, P.F., Cannell, G.H., Garber, M.J. and Bair, F.L. (1967) Effect of three nitrogen fertilizers on gains, losses and distribution of various elements in irrigated lysimeters. *Hilgardia* **38**: 265-283

Quirk, J.P. and Schofield, R.K. (1955) The effect of electrolyte concentration on soil permeability. *J. Soil Sci.* **6**: 163-178

Rawlins, S.L. and Raats, P.A.C. (1975) Prospects for high-frequency irrigation. *Science* **188**: 604-610

Reeve, R.C. and Bower, C.A. (1960) Use of high-salt waters as a flocculent and source of divalent cations for reclaiming sodic soils. *Soil Sci.* **90**: 139-144

Reeve, R.C. and Doering, E.J. (1966) The high-salt dilution method for reclaiming sodic soils. *Soil Sci. Soc. Amer. Proc.* **30**: 498-504

Reeve, R.C., Pillsbury, A.F. and Wilcox, L.V. (1955) Reclamation of a saline and high boron soil in the Coachella Valley of California. *Hilgardia* **24**: 69-91

Rhoades, J.D. (1968) Leaching requirement for exchangeable-sodium control. *Soil Sci. Soc. Amer. Proc.* **32**: 652-656

Rhoades, J.D. (1972) Quality of water for irrigation. *Soil Sci.* **113**: 277-284

Rhoades, J.D. (1974) Drainage for salinity control. Chapter 16 in: van Schilfgaarde, Jan (ed.), Drainage for Agriculture. *Agronomy 17, Amer. Soc. Agron.*, Madison, Wisconsin

Rhoades, J.D. (1975) Measuring, mapping and monitoring field salinity and water table depths with salt resistance measurements. *Proc. of Expert Consultations on the Prognosis of Salt-Affected Soils, Rome;* June 3-6. FAO, United Nations, Rome, Italy

Rhoades, J.D. and Ingvalson, R.D. (1971) Determining salinity in field soils with soil resistance measurements. *Soil Sci. Soc. Amer. Proc.* **35**: 54-60

Rhoades, J.D., Oster, J.D., Ingvalson, R.D., Tucker, J.M. and Clark, M. (1974) Minimizing the salt burdens of irrigation drainage waters. *J. Environ. Qual.* **3**: 311-316

Richards, L.A. (ed.) (1954) Diagnosis and improvement of saline and alkali soils. U.S. Dept. of Agriculture, Handbook 60, 160 pp

Shainberg, I. and Otoh, H. (1968) Size and shape of montmorillonite particles saturated with Na/Ca ions. *Israel J. Chem.* **6**: 251-259

Shainberg, I. (1973) Ion exchange properties of irrigated soils. Chapter IV.1 in: Yaron, B., Danfors, E. and Vaadia, Y. (eds.), Arid Zone Irrigation. Springer Verlag, Berlin, 433 pp

Shalhevet, J. (1970) The use of saline water for irrigation. In: Dialogue in Development, 2nd World Congress of Engineers and Architects. Tel Aviv. Dieter Gerson (ed.), pp. 437-442

Shalhevet, J. (1973) Irrigation with saline water. Chapter VI.1 in: Yaron, B., Danfors, E. and Vaadia, Y. (eds.), Arid Zone Irrigation. Springer Verlag, Berlin, 433 pp

Shalhevet, J and Bernstein, L. (1968) Effects of vertically heterogeneous soil salinity on plant growth and water uptake. *Soil Sci.* **106**: 85-93

Talsma, T. (1967) Leaching of tile-drained saline soils. *Aust. J. Soil Res.* **5**: 37-46

van Schilfgaarde, Jan, Bernstein, L., Rhoades, J.D. and Rawlins, S.L. (1974) Irrigation management for salt control. *J. Irrig. and Drainage Div., ASCE* **100 (IR3)**: 321-338

Wadleigh, C.H. and Ayers, A.D. (1945) Growth and biological composition of bean plants as conditioned by soil moisture tension and salt concentration. *Plant Physiol.* **20**: 106-132

Wilcox, L.V. (1960) Boron injury to plants. *U.S. Dept. Agr. Inf. Bull.* **211**, 7 pp

Yaron, B., Danfors, E. and Vaadia, Y. (eds.) (1973) Arid Zone Irrigation, Springer Verlag, Berlin. 433 pp

Yaron, B., Shimshi, D. and Shalhevet, J. (1973) Patterns of salt distribution under trickle irrigation. In: Physical Aspects of Soil Water and Salt in Ecosystems. *Ecological Studies* **4**: 389-394. Springer Verlag, Berlin